OLYMPIC STADIA

OLYMPIC STADIA provides a comprehensive account of the development of stadia including but not limited to: developments in running tracks, the introduction of lighting, improvements in spectator viewing standards and the introduction of roofs.

Written by a world-renowned expert on sports architecture and an award-winning engineering journalist, the book:

- Systematically analyses every stadium from Athens 1896 to Tokyo 2020
- Provides drawings, plans, elevations, photographs and illustrations in full colour
- Considers the fundamental changes wrought by the incorporation of the Paralympic Games
- Looks at the impact on host cities and their urban infrastructure, and considers the long-term legacies and massive investments that Olympic stadia require
- Explores the effects of the demands of the world's TV broadcasters.

An invaluable and beautiful resource for practical insight and inspiration, this book makes essential reading for anyone interested in Olympic stadia.

GERAINT JOHN, Dip. Arch. (UCL), RIBA, Companion CIMPSA, FRSA, has a deep involvement in buildings for sport and leisure. His previous experience as the chief architect and head of the Technical Unit for Sport at the GB Sports Council has made him an expert in the particular field of sports facilities. He is a joint author of *Stadia: The Populous Design and Development Guide*, now in its fifth edition, and has curated exhibitions at the Olympic Museum in Lausanne.

In 2014, the International Olympic Committee awarded Geraint the Pierre de Coubertin Medal for outstanding services to the Olympic movement. He is the only British subject to be so honoured.

He is a senior advisor to Populous, a global architecture and design firm.

Geraint is a Visiting Professor at the University of Hertfordshire. He sat on the Environment Committee of the London bid for the 2012 Olympics and on the government's Global Sports Projects Sector Advisory Group.

He is Honorary Life President of the International Union of Architects' Sports and Leisure Group, and an inaugural member of the International Association of Sports Facilities Hall of Fame.

DAVE PARKER B.Sc., CEng., FICE, was technical editor of *New Civil Engineer* magazine for 14 years before leaving in May 2006 to become a freelance author and journalist. *New Civil Engineer* is published by Emap under licence from the Institution of Civil Engineers, and goes monthly to all members of the ICE.

Prior to becoming a journalist 30 years ago, Dave was a practising civil engineer for more than 25 years. He worked in both design and contracting, was senior partner of a small forensic engineering practice, and technical marketing director of a specialist products company.

Dave is also a former Visiting Professor of Civil Engineering at the Queen's University of Belfast, a position he held for ten years.

In 2014 Dave was asked to return to *New Civil Engineer* as technical editor emeritus.

OLYMPIC STADIA

THEATRES OF DREAMS

GERAINT JOHN and DAVE PARKER

Routledge
Taylor & Francis Group

LONDON AND NEW YORK

First published 2020
by Routledge
2 Park Square, Milton Park, Abingdon, Oxon OX14 4RN

and by Routledge
605 Third Avenue, New York, NY 10017

Routledge is an imprint of the Taylor & Francis Group, an informa business

British Library Cataloguing-in-Publication Data
A catalogue record for this book is available from the British Library

Library of Congress Cataloging-in-Publication Data
Names: John, Geraint, author. | Parker, Dave (Journalist), author.
Title: Olympic stadia : theatres of dreams / Geraint John and Dave Parker.
Description: New York : Routledge, 2020. | Includes bibliographical references
    and index.
Identifiers: LCCN 2019021000| ISBN 9781138698840 (hb : alk. paper) |
    ISBN 9781315518053 (ebook : alk. paper)
Subjects: LCSH: Olympics—History. | Stadiums—History.
Classification: LCC GV721.9 .J64 2020 | DDC 796.4809—dc23
LC record available at https://lccn.loc.gov/2019021000

Typeset by Alex Lazarou

ISBN 13: 978-1-138-69884-0 (hbk)

CONTENTS

To my family and Jennifer Purvis for their help and support during a period of health and family problems.

GERAINT JOHN

To my late wife Lesley, who battled cancer throughout the creation of this book, but did not live to see its completion. And to my daughter Jenny, whose support to us both made it all possible.

DAVE PARKER

ACKNOWLEDGEMENTS

I MUST acknowledge with great thanks all the contributions that have been made in the preparation of this book. I also want to pay tribute to those who have had an influence on my professional life in the field of the architecture and development of buildings for sport and leisure.

I will begin with Rod Sheard and Bill Slater, two people who have supported me and have been of major influence, and Professor Llewelyn Davies, who introduced me to the concept of the architect as Renaissance Man.

I would like to particularly thank those who have helped in the preparation of this book: Kengo Kuma, Soichiro Sano, Jac Griffiths, Sena Ofliz, Damon Lavelle, Benjamin Flowers, Li Xinggang, Arthur Gelling, Paul Rushton, Gustavo do Amarol and Paul V. Dudman. The names of Shane Kane, Trevor Ruddell and Gilson Santos should be added.

The practice of Populous has been an enormous support to me. Damon Lavelle, Tom Jones and Ben Vickery are worthy of special mention related to this project, together with Christopher Lee and Nicholas Reynolds. Matt Grace, Maria Vicalova, Jess Lockhart and Paul Shakespeare must also be included.

Those who have stimulated me throughout my career include Ian McKenzie, Bill Stonor, Simon and Jackie Inglis, Derek Wilson, Professor Terry Stevens, Dyfed Griffiths, Stephen Brookhouse, Silvio Carta and Professor Peter Cook.

I would like to acknowledge the help I have received from colleagues in the International Olympic Committee (IOC): Sir Craig Reedie, Gilbert Felli, Christopher Dubi, Anna Volz Gott, Michelle Lemaitre, Anne Chevalley, Alain Quenzer, Rachel Caloz, the staff of the Olympic Studies Centre, and from the International Association for Sports and Leisure Facilities (IAKS), Klaus Meinel.

Through the International Union of Architects (UIA) I want to mention those who have stood out as colleagues who have supported and inspired me: Wojciech Zablocki, Gar Holohan, Kresimir Ivanis, Pino and Alessandro Zoppini, David Body, Augustin Garcia Puga, Christophe Parade, Jorge Joppien, Nikos Fintikakis, Rene Kural, Joaquin Pujol, and Harald Fux. I should add the many others too numerous to mention here.

From those whose names I have missed, I ask forgiveness. There are many.

GERAINT JOHN
July 2018

MANY RELIABLE sources are quoted in this book – but there are many others that turned out to be inaccurate, incomplete or simply misleading. For most of the stadia described herein the primary and most trusted sources were the Official Reports produced by the local Olympics Committees and available on the Olympic.org website.

There was one exception. After the Antwerp Summer Games of 1920 the local committee went bankrupt. Nothing comparable to the other Official Reports was ever produced, and detailed and reliable information on the stadium was remarkably hard to find.

I must therefore record my gratitude to Hertfordshire Libraries and particularly the staff of Bishop's Stortford library, who finally managed to track down a copy of a definitive Belgian book on the 1920 Games, in Canada! This proved to be invaluable, as Chapter 10 will, I hope, illustrate.

Special thanks also to "Jen's Red Pen", my daughter Jenny's proofreading service.

DAVE PARKER
August 2018

FOREWORD
LORD COE

AT THE HEART of every Olympic Games is the Olympic stadium, the epicentre around which all events revolve and where history is made.

Olympic stadia are now defining venues in their own right. They are unique and complex structures that must keep pace with technology, social change and environmental needs to ensure their own sustainability. They shape and influence the urban landscape and character of a city and have provided, for spectators worldwide, the backdrop for captivating opening and closing ceremonies and moments of exceptional sporting achievement.

This book is not just about the history of Olympic stadia. It also highlights the lessons learned from the evolution of the design and construction of these significant landmarks.

The chronicle of how Olympic stadia have evolved and consideration of the increasing demands for the future make this book a valuable reference for architects and design teams, particularly from cities bidding to host the Olympic Games.

INTRODUCTION

BY THE TIME the Tokyo 2020 Opening Ceremony takes place there will have been a grand total of 27 stadia that have hosted the Summer Games of the modern era. Of these, 14 were new, built specially for the Games; 13 were reconstructions, extensions or upgrades of existing stadia. Only one equivalent stadium has ever been built for the Opening and Closing Ceremonies of the Winter Games: Sochi's Fisht Stadium. This book will cover the architecture and structural design of each stadium. It will also set each in its social, political, and economic context, and consider its long-term legacy.

In the 19th century the early attempts to revive the Olympics were inspired by what was then known – or thought to be known – about the Ancient Greek Games. How these beliefs influenced the development of the modern Summer Games in particular has to be considered. Ancient events such as the standing long jump may have been abandoned after the 1912 Summer Games, but sports such as wrestling, the javelin and discus, archery and foot races can trace their origins back to Ancient Greece, as can the whole concept of a dedicated stadium.

A much later major influence on the evolution of stadium design was the arrival of the Paralympics after the Second World War, whose beginnings were inspired by Sir Ludwig Guttmann.

By the 21st century the Summer Games had become the biggest and most important spectacular sports events in the world. Every four years live satellite television transmissions transform billions of people worldwide into participating spectators. It is said that 70 per cent of the world's population watched the 2008 Olympics in Beijing.

Many of the Ancient Greek Olympic events and traditions are still relevant to the modern Summer Games.
SOURCE: CREATIVE COMMONS – CREDIT LIVIOANDRONICUS2013

Each successive Olympics is likely to be the biggest media and broadcasting event ever. It is difficult to be precise about numbers, but in London 2012 the estimated attendance of media representatives was 28,000, a figure many times greater than that of the athletes competing.

In London, a "high street" was built for the media which contained a pharmacist and technology shop, and a fully kitted out health centre staffed by doctors and nurses. Such is the growing importance given to the media at these events.

At the first Summer Games of the modern era in Athens in 1896 around 240 men competed in nine sports. At Tokyo in 2020 close to 5,000 men and an equal number of women will compete in 324 events in 33 sports. All these will require an appropriate venue.

The Olympic Games are indissolubly linked to the city that hosts them. These host cities are now transformed by the building of the facilities, and by the infrastructure that is needed to service this gigantic event. Early Games infrastructure was designed with little thought to its legacy – now legacy is one of the most important design parameters of all.

At the heart of almost every Summer Games will be the main stadium. There are two exceptions to this over the decades: Paris 1900, and Rio 2016. Usually it is the largest venue with the greatest spectator capacity. It will host the Opening and Closing Ceremonies, ensuring it will be for a few short weeks the most watched, most famous building in the world.

Spectacle is now at the heart of the Opening Ceremony. Seoul's Olympic Cauldron at the 1988 Summer Games was an outstanding example.
SOURCE: CREATIVE COMMONS – CREDIT KEN HACKMAN, USAF

It has been called "a Temple of Sporting Spectacle" equipped with the most modern technologies. It can be a magical place, where legends are born.

At its best, it is not just a stage for sporting events. It is a huge theatre for the exhibition and performance of historic feats.

- It can be a connection between sport and its urban surroundings. It is a unique opportunity to design a building that can incorporate and respect the history and culture of the city.
- It can help to influence the renewal of neighbourhoods, and regenerate depressed areas. It should be part of a development plan that integrates the stadium into an overall concept that will benefit the community, the city, the region and host country. Properly designed, the stadium can become an important multi-purpose entertainment complex for the city after the Games are over.
- It can open up new avenues of development in structural design, architecture and materials technology.
- Its part in the legacy of the Games will be hugely important.
- Though built for an event of limited duration, its existence will be long lasting.
- Its dramatic function and often monumental scale can combine to produce powerful civic architecture and structures.
- At its best, it should be accessible to and loved by the public.

History has examples of both success and failure. Some stadia have stood the test of time and are still functioning for the city. They have probably been extensively modernized and upgraded during their lifetime. Others are little used, perhaps due to a weak legacy or a failure to adapt.

Atlanta's Turner Field was once the short-lived Centennial Olympic Stadium, main venue for the 1996 Summer Games, which was specifically designed for legacy use as a baseball park.
SOURCE: CREATIVE COMMONS – CREDIT ZPB52

One perpetual problem is the reality that most popular spectator sports, such as soccer, American football and the like, fit uncomfortably into a basically oval-shaped stadium. Only cricket and Australian Rules football are at home in such a venue. Adapting such a stadium to serve as a baseball park is an even greater challenge, one that only Atlanta came close to meeting.

And what of the future? How many more new stadia will be built to host future Games? How many countries will be prepared to fund ever more futuristic, complex and expensive venues and their ever more spectacular Opening and Closing Ceremonies? Where once up to 16 cities would compete for the honour of hosting the Summer Games, now it is hard to attract realistic bids from more than two or three, usually based on the re-use of existing facilities.

And yet, there have been previous periods in Olympic history when the Games faced remarkably similar problems. There was only one bidder for the 1984 Summer Games, for example. Yet the Games survived, and competition to stage them intensified, and some spectacular, iconic stadia were created.

In the future, Olympic stadia, be they greenfield projects or rebooted existing venues, will be designed with the benefits of many decades of experience. Sustainability and care

for the environment will be an absolutely key parameter. The "overlay" that surrounds the main stadium – the temporary buildings that house the media and the security personnel, the warm-up facilities, the canteens and restaurants for the athletes, officials and media – should be an integral part of the stadium planning from the outset.

Creating a successful stadium with a meaningful legacy is a major challenge. What is clear is that the best design talents must be brought together for its design.

In 1910 Baron Pierre de Coubertin, whose vision launched the modern Olympics movement, wrote in the *Olympic Review*: "It is for the Architects now to fulfil the great dream, to let soar from their imagination a resplendent Olympia, at once original in its modernism and imposing in its traditionalism, but above all perfectly suited to its function. And who knows? Perhaps the hour will strike when the dream already committed to paper, will be built in reality."

Has the hour ever struck? Many dreams have become solid reality over the decades, as later chapters will describe in detail. Many design teams have been inspired by de Coubertin's words. It is to be hoped that designers of future Olympic stadia will also take similar inspiration, and strive to create a stadium that would receive de Coubertin's wholehearted approval.

The only purpose-built Winter Olympics stadium was host to a spectacular Opening Ceremony for the 2014 Winter Games in Sochi.
SOURCE: CREATIVE COMMONS – CREDIT KOREAN CULTURE AND INFORMATION SERVICE

OLYMPIC STADIA
THEATRES OF DREAMS

CHAPTER 1

IN THE BEGINNING

BY THE YEAR 200 BCE the ancient Olympic Games had reached their classical apogee. Every four years a truce was agreed between the frequently warring Greek city states, a truce largely observed, and hundreds of athletes from all over the Greek world made the difficult, uncomfortable and expensive journey to the city of Elis on the Peloponnese peninsula in southern Greece. Nearby was a complex of marble temples and monuments, and a dedicated athletics stadium. This was ancient Olympia, a name that had no direct connection to Mount Olympus, far to the north.

Up to 20 competitions were held over seven days, usually including around six or seven equestrian events such as chariot racing. The prizes for the winners would be no more than olive leaf wreaths – although those who were crowned with the olives could expect to be rewarded with money, gifts and even political advancement when they returned to their home cities.

There were other Games in Greece at the time; in fact most major cities held their own Games, but the most prestigious were the Panhellenic Games. These were held at Olympia, Delphi, Nemea and Isthmia, with the Olympics being the longest established and most revered.

An early artist's impression of ancient Olympia, not confirmed by subsequent archaeological investigations.
SOURCE: PUBLIC DOMAIN

In practice this meant that at least one of these four Games was held every year. Together they were also known as the Stephanitic Games, meaning that the victors received only a wreath, as opposed to the Panathenaic Games, for example, where winners could expect large quantities of olive oil as prizes.

All four Games were as much religious festivals as sporting events. Each of the Games honoured a different Greek god: Zeus in the case of the Olympics, whose prestige had much to do with a giant 13m high gold and ivory statue of Zeus that was considered one of the seven wonders of the ancient world. For the first 13 Olympiads only one event took place, the stadion, a simple foot race over 192m, and this was seen as a relatively minor feature of the elaborate religious ceremonies.

Indeed, the first recorded race at the Olympics, in 775 BCE, was a stadion for women, to decide who would be priestess for the goddess Hera. Later the Heraean Games for women were held from around 600 BCE. Women competed, in men's clothing, in a variety of foot races.

As male competitors were nude, it was and is widely believed that women were banned from watching the Games on pain of being thrown off a nearby cliff. It now seems, however, that this ban only applied to married women.

Women also competed in their own Games, wearing male clothing.
SOURCE: PUBLIC DOMAIN

Competitions for youths were first recorded at the 37th Olympiad in 632 BCE. Many of the 23 events that took place when the Games were at their height have never been revived. These included mule-cart racing, the virtually rule-free combat sport pankration and the final event at each Games, the hoplitodromos, where competitors raced in full military armour.

However, boxing, wrestling and a range of foot races have survived. The long jump, the javelin and the discus, along with wrestling and a foot race, formed the first pentathlon.

After the Roman conquest of Greece in 148 BCE the Games continued for several centuries. A gradual decline led to gladiatorial combats and wild animal hunts being added to pander to the more robust Roman taste. Earthquakes and barbarian invasions,

A classic image of ancient Greek athletes.
SOURCE: CREATIVE COMMONS – CREDIT MARIE-LAN NGUYEN

plundering by the Romans and frequent flooding all took their toll on the Olympic complex. The death knell was sounded in AD 394, when Theodosius, the Christian Emperor of Rome, banned all Games throughout the Greek world as being no more than pagan festivals.

Remarkably, memories of the Olympics lingered on over the centuries, never quite fading away. In the 19th century there were serious attempts to revive the Olympic spirit (see Chapter 2), culminating in the first Olympiad of the modern era (see Chapter 5).

Further reading
Swaddling, Judith (2004) *The Ancient Olympic Games*, The British Museum Press, ISBN 978–0–7141–2250–2.

The original Olympic stadium, set in its natural amphitheatre.
CREDIT: PUBIC DOMAIN

CHAPTER 2

FIRST STIRRINGS

IMPERIAL ROME may have first corrupted and finally crushed the Ancient Greek Games (see Chapter 1), but the Olympic Flame never quite flickered and died. For millennia human beings have congregated, celebrated and competed. Religious festivals, funeral ceremonies and solstices have provided many opportunities, and even when the doctrinal heirs of the Christian Roman emperors tried to suppress such "pagan" events, they never completely succeeded.

As Rome's amphitheatres and arenas decayed and largely disappeared, such events as were held often returned to their ancient roots in natural arenas. A prime example is located on Dover's Hill, on the outskirts of Chipping Campden in the Cotswolds, an area in the south-west of England. In the early 17th century the area had grown rich from the wool trade, but political and religious strife was convulsing the country. A local Cambridge-educated solicitor by the name of Robert Dover took on the task of unifying local residents, if only for a day or two.

Whitsuntide, the week following Whit Sunday, itself seven days after Easter, was a traditional holiday period and the time for fairs, processions and parades. Dover proposed a series of competitions that would attract competitors and spectators from all social classes, an event that was soon to be christened the Cotswold Olimpicks.

It was held in a natural amphitheatre on the side of what would later be named Dover's Hill in tribute to Robert Dover's inspiration. A steep bank on one side allowed spectators a good view of the arena, and there were wonderful views over the Vale of Evesham. The first Games were held in the early years of the 17th century, probably around 1612.

Few of the events that took place bore much resemblance to Ancient Greek sports. There was running, jumping and wrestling, and sledgehammer throwing, but also horse racing, coursing with hounds, dancing competitions, quarterstaff bouts, and fighting with swords and cudgels.

A temporary timber building dubbed Dover's Castle was mounted with small cannon, which signalled the start of each event. There were also booths and pavilions where chess and various card games were played, and plentiful food was supplied.

Over the years the Games became increasingly popular, attracting spectators from all social classes from as far away as London (a journey of at least three days), including even royalty, in the personage of Prince Rupert of the Rhine, a nephew of Charles I. However, the Games were terminated when the English Civil War broke out in 1642.

Revived after the Restoration of the monarchy in 1660, the Games continued until 1852, when the common land on which they were held was seized by local landowners. By that time they were said to have degenerated into a drunken and disorderly country festival, a view that might well have helped the local landowners get their Inclosure Act (sic) through Parliament.

Few of the events that took place at the Cotswold Olimpicks bore much resemblance to Ancient Greek sports – and it is doubtful if Dover's Castle was ever this elaborate.
SOURCE: PUBLIC DOMAIN

The earliest image of the Wenlock Olympian Games with Windmill Hill in the background, June 1867.
SOURCE: Wenlock Olympian Society

Dr William Penny Brookes is said to be the father of the modern Olympic Games, after he staged the first Wenlock Olympian Games in 1850.

SOURCE: Wenlock Olympian Society

Not far to the north, however, a new Olympic seed was taking root. In the county of Shropshire, in the small town of Much Wenlock, local doctor William Penny Brookes was leading a movement to improve the moral, physical and intellectual capacity of local inhabitants, especially the working classes. In February 1850 the Wenlock Agricultural Reading Society he established voted to set up the "Olympian Class", with Brookes as its secretary, and in October of the same year the first Games were held at Much Wenlock racecourse.

Events included athletics, football and cricket, and penny-farthing cycle racing, along with the traditional country sport of quoits. Some early Games also staged fun events such as blindfold wheelbarrow racing and an "old woman's race", where the winner was awarded a pound of tea.

A prize-giving at the Wenlock Olympian Games, with Dr Brookes and his medals well to the fore.

SOURCE: Wenlock Olympian Society

In 1859 the Class sent £10 to Athens as the prize for the victor in the Long Foot Race at the first Zappas Olympics (see Chapter 5), and in 1860 the name was changed to the Wenlock Olympian Society. Brookes did not rest there, however.

Thanks to his efforts, the first Shropshire Olympian Games were held in 1861. These would move from town to town throughout the county. Three years later the Society awarded a silver medal to a certain John Hulley of Liverpool, and made him an honorary member. This was in recognition of his organization of the Mersey Olympics, described as a "Grand Olympic Festival" and also known as the Liverpool Olympics, which was first staged in 1862.

Hulley was a typical Victorian entrepreneur with a keen interest in sports, particularly gymnastics and athletics. Some 10,000 spectators are said to have attended the first Liverpool Olympics: so successful were they deemed to be that five more events were staged in subsequent years. Four different locations were used: Liverpool's Mount Vernon Parade Ground, the Zoological Gardens, the Sheil Park Athletic Grounds, and the Llandudno Croquet Grounds in Wales, twice.

Brookes and Hulley, along with a representative of London's German Gymnastic Club, formed the National Olympian Association (NOA) in 1865. The first National Olympian Games were held at the Crystal Palace, London, the following year. More than 200 athletes from all social classes competed in athletics, gymnastics, swimming, boxing, wrestling and fencing. Again, attendance was high, topping the 10,000 mark.

Birmingham was chosen as the location of the second Olympian Games in 1867. The three-day event was again hailed as a great success, and Manchester was selected as the next host city. Unfortunately, a massive dispute between the NOA and the newly formed Amateur Athletic Club, which defined amateur as "gentleman amateur", and problems with the venue, saw a much-reduced event held in Wellington, Shropshire, in 1868. This spelled the beginning of the end for the NOA, which was terminated in 1883.

Early attempts to revive the Ancient Olympics were not confined to Britain and Greece. In France, during the French Revolution, there were three games held between 1796 and 1798. These had the distinction of being the first occasions in which runners competed over distances defined in metres, Revolutionary France having just instituted the metric system. Eighty-five years later, a young French aristocrat, heavily influenced by the English public school system, was unsuccessfully campaigning for a greater emphasis on physical education in French schools. Thwarted, Baron Pierre de Coubertin turned to campaigning for the revival of the Ancient Games, a mission that ultimately succeeded in 1894 with the formation of the International Olympic Committee (see Chapter 3).

His main inspiration was the success of the Wenlock Games. In 1889 de Coubertin received a letter from Brookes inviting him to visit Shropshire and discuss the Olympic campaign. De Coubertin arrived in 1890, and stayed in Brookes' family home for several days. Brookes had organized a staging of the Wenlock Games in his honour, with elaborate pageantry, followed by a lavish dinner in a local hotel.

In their discussions, Brookes would have informed de Coubertin of his attempts to internationalize the Olympic revival movement by corresponding with the organizers of the first Zappas Olympics and, subsequently, with the Greek government. Despite his efforts, the Zappas Olympics were restricted to Greek-speaking competitors.

De Coubertin must have learned a great deal from Brookes' experience with the various English Games, including the advantages of an opening pageant, the importance of amateurism and inclusion, and the benefits from staging such events at different locations. These included the crucial information that the host cities for the National Olympian Games funded the events, not the organizing committee.

At the time, de Coubertin expressed immense gratitude to Brookes. In later years he tended to play Brookes' contribution down, but there are still obvious links between these early Olympic stirrings and the massive global razzmatazz that is the modern Olympic Games.

Brookes died in December 1895 at the age of 84, four months before the first modern Olympic Games in Athens. His legacy lives on. The Wenlock Olympian Games are still held annually, as are the Cotswold Olimpicks. In 1994 a wreath was laid on his grave in Much Wenlock churchyard by the then President of the IOC, Juan Antonio Samaranch.

Samaranch said: "I came to pay homage and tribute to Dr Brookes, who really was the founder of the modern Olympic Games."

CHAPTER 3

BARON PIERRE DE COUBERTIN AND THE BIRTH OF THE MODERN GAMES

IN 1894, 79 delegates from 12 countries met at the Sorbonne in Paris. This congress was the culmination of one man's campaign to revive the Ancient Greek Olympic Games. The man was a 31-year-old French aristocrat, historian and educational pioneer, Baron Pierre de Coubertin, and the outcome was the establishment of the International Olympic Committee (IOC), and the allocation of the first Summer Games to Athens, Greece, and the second to Paris.

Although de Coubertin is often dubbed "the father of the modern Olympics", the 1894 congress was far from the first attempt to bring back the Ancient Games. There had been earlier revivals, in both England and Greece, and also in Revolutionary France (see Chapter 2). De Coubertin's vital and seminal contribution was the concept of an international Games, held in a different city in a different country every four years. Previous "Olympics" had been purely local events – de Coubertin's vision was of teams of international athletes competing against each other for the glory of their country.

Baron de Coubertin's vision of international Games held in a different city every four years was the spark that ignited the modern Olympics movement.
SOURCE: CREATIVE COMMONS – CREDIT DUTCH NATIONAL ARCHIVES

Born in Paris in January 1863 into a wealthy aristocratic family, de Coubertin grew up in an era of profound political and social change in France. He was just 8 years old when France crashed to humiliating defeat in the Franco-Prussian War (1870–1871). Emperor Napoleon III was driven into exile, civil war broke out, and political turmoil continued for decades.

Later, de Coubertin became convinced that one key reason for the French Army's generally woeful performance in the war was its poor physical condition. A deep interest in education led to an 1883 study of schools in Great Britain and their programmes of physical education. De Coubertin was particularly impressed by the classic public schools, including Rugby under its famous headmaster Thomas Arnold.

His own education was rigorous and moralistic. Sent to a Jesuit boarding school in Vienna, de Coubertin proved to be an outstanding student. From these schooldays and his research in England came his first major campaign: for a much greater emphasis on physical education in French schools. Organized sport encouraged both physical and intellectual development, he believed.

Despite his social status and great wealth, de Coubertin's first campaign failed. Undaunted, he threw all his energies, political skills and, to a certain extent, his wealth, into a new, even more ambitious campaign – the revival of the Games. As a historian he had been fascinated by Ancient Greece for many years, particularly the Athenian Gymnasia. These encouraged young men to develop both their physiques and their intellectual capacity, a concept that de Coubertin was to pursue for much of his adult life.

A meeting with Dr William Penny Brookes, creator of the Wenlock Olympian Games (see Chapter 2), was a seminal influence on de Coubertin. He began a five-year campaign, which initially yielded some applause, but soon ran into parochial lack of interest from French sport administrators.

In fact the hot topic of the day was amateurism in sport. Betting syndicates and sponsors were beginning to destroy the traditional amateurism. De Coubertin had helped found a national association to coordinate athletics in France: now, under its banner, he organized a congress ostensibly to discuss amateurism.

This plaque commemorates the planting of an oak tree by de Coubertin during his visit to Much Wenlock in 1890. The tree continues to thrive.

When the delegates assembled on 23 June 1894, they were promptly divided into two commissions: one to discuss the future of amateurism, the second to consider de Coubertin's proposal to revive the Olympics. Somewhat predictably, the amateurism commission produced little of substance, but it did set down some important principles for the Games, such as the use of heats and the banning of most forms of prize money.

De Coubertin, naturally, headed up the second commission. He delivered a rousing speech, convincing the experts and guests present to support the setting up of the International Olympic Committee (IOC) and the revival of the Ancient Games. It was also agreed that the Summer

Games would be held every four years, and that the programme be of modern rather than ancient sports.

An Olympic motto was proposed by de Coubertin – "Citius, Altius, Fortius" (Faster, Higher, Stronger; originally coined by his friend Henri Didon, a Dominican monk and athletics enthusiast) – but he had to wait until 1924 to see it officially adopted.

Ancient Greece may have inspired de Coubertin and his supporters, but it was a highly romanticized 19th-century version of Greek history that shaped the development of the Olympic movement (see Chapter 1). There were to be no female competitors, if de Coubertin had his way, especially not in athletics. There would be no money prizes or professional athletes – although professional fencing masters were allowed to compete.

From its beginnings the platform of the IOC was "Sport and Culture" (the environment was added later). De Coubertin strongly believed that organized sport was an agent of both physical and cultural renewal. He was convinced that alongside the sporting competitions there should be artistic competitions of equal stature, and achieved his aim at the 1912 Stockholm Summer Games.

There, gold medals were awarded for literature, sculpture, painting, architecture and music. De Coubertin entered a poem – under an assumed name – and duly won first prize in the literature competition. These competitions continued until 1948, only to be discontinued when it was ruled that the artists competing were professionals rather than amateurs.

Ancient Greek athletes were seen by most as pure amateurs, striving for the glory of the winner's wreath rather than any pecuniary reward. De Coubertin's beliefs on how this should be recreated for the modern Games were somewhat mixed, in that, while he advocated strict rules against professionalism, he later came to believe that working-class competitors should be compensated for wages lost during the Games.

He also held firm to the belief that participation was more important than victory and that the internationalist ethos underlying the Games could give them an important role in promoting world peace.

Rugby football was another of de Coubertin's passions. In 1892 he refereed the first ever French Championship final: in 2007 he was inducted into the International Rugby Board Hall of Fame.

Pierre de Coubertin died in Geneva in September 1937 and was buried in Lausanne. On his instructions, his heart was removed and placed in a marble stele commemorating the revival of the Olympic Games, close to the ruins of Ancient Olympia.

Baron de Coubertin's heart was interred in this marble stele close to the site of Ancient Olympia.
SOURCE: CREATIVE COMMONS – CREDIT TROY MCKASKLE

In 1964 the IOC inaugurated the Pierre de Coubertin Medal, the highest honour it can award. Professor Geraint John, co-author of this book, was awarded the Medal in November 2014 for outstanding services to the Olympic movement.

The Pierre de Coubertin medal presented to co-author Professor Geraint John.
CREDIT: GERAINT JOHN

CHAPTER 4

SIR LUDWIG GUTTMANN AND THE BIRTH OF THE PARALYMPIC GAMES

BEFORE THE 1940s the outlook for patients with serious spinal injuries was grim. Average survival time after the injury was just six weeks. Those who survived longer, who avoided untreatable bladder and kidney infections, faced a miserable future – either bedbound and slowly wasting away or confined to primitive wheelchairs. Hidden away in grim institutions, they had effectively been written off by society.

Two seminal events in 1944 began a revolution in patient care. Penicillin, the first effective antibiotic, went into full-scale production in the USA. And in England, a refugee German Jewish neurosurgeon was appointed director of the new Stoke Mandeville Spinal Injuries Unit, located less than 80km (50 miles) north-west of London.

Both events were a response to the mass casualties expected during and after the D-Day landings in Europe. Penicillin ensured that far fewer military personnel with severe spinal injuries succumbed to the infections

Neurologist Sir Ludwig Guttmann (1899–1980) is regarded as the founder of the Paralympic Games.
CREDIT: COURTESY OF WHEELPOWER/STOKE MANDEVILLE STADIUM.

that killed so many in earlier years. Penicillin on its own, however, would have only increased the number of survivors enduring a primitive and ineffective long-term treatment routine. What was needed was a completely new approach to patient care for the severely disabled – and it was pioneered in a 12-bed converted Nissen hut in the grounds of Stoke Mandeville Hospital.

Leading the revolution was Dr Ludwig Guttmann, who had fled to Britain in 1939 to escape Nazi persecution. By all accounts he was a formidable and autocratic personality, unshakeable in his conviction that he was doing the right things, impervious to the horrified initial reactions from fellow doctors, nurses and even the patients. He persevered with such controversial measures as regularly turning bed-bound patients every few hours, setting up workshops where patients could learn valuable skills and trades – and introducing sporting activities into the treatment of wheelchair users. As his treatment plans began to show dramatic improvements the objections melted away.

Guttmann was born in 1899, in what is now the Polish city of Toszek. After qualifying in 1924 he specialized in neurosurgery and within nine years he was regarded as the leading neurosurgeon in Germany. When the Nazis seized power in 1933 they soon banned Jews from practising medicine outside hospitals for Jews. Guttmann could continue to work at the Jewish hospital in Breslau (now Wroclaw in Poland), but the violent attacks on Jews and Jewish properties during the so-called Kristallnacht in November 1938 convinced him that he had to escape from Germany.

All Jewish passports had already been confiscated. Late in 1938, however, a close friend of the Portuguese dictator Salazar fell seriously ill with a rare neurological condition. To improve their relations with Salazar the Nazis offered Guttmann's services, and the offer was promptly accepted.

Guttmann was due to return from Lisbon via London. He had already been in contact with the Society for the Protection of Science and Learning, which campaigned to protect persecuted scholars. Once in London, he got the news that visas had been arranged for him and his family and were waiting for him in Berlin.

Together with his wife and two small children Guttmann arrived in England in March 1939. The Council for Assisting Refugee Academics provided funding, and he and his family settled in Oxford, living for a time as guests of the Master of Balliol College. Guttmann continued his research into spinal injuries at the city's Radcliffe Infirmary, until the British government asked him to establish the Stoke Mandeville facility.

At this time Stoke Mandeville was for military patients only; young men frustrated and depressed by their loss of mobility. Guttmann dreamed up new sports they could compete in: wheelchair archery, table tennis and basketball to begin with. (Trials of wheelchair polo were discontinued after the combination of walking sticks and a hockey puck proved too dangerous.)

It was not just the benefits of exercise the patients enjoyed. Spirits were lifted, boredom dispelled, long-term friendships forged. The London Olympics of 1948 gave Guttmann fresh inspiration.

In parallel with the London event he organized the first Stoke Mandeville Wheelchair Games. This seminal event was an almost unnoticed archery competition for 16 war veterans, including two women. For the next three years the Games were a purely local British affair, then, in 1952, Dutch and Israeli teams of wheelchair users participated, leading the event to be dubbed the 1st International Stoke Mandeville Games.

From then until 1995 the Games were held annually, after which it was rechristened the World Wheelchair Games, subsequently undergoing a number of name changes but still continuing on an annual basis at international venues. The most transformational of these were the 9th Games in Rome in 1960.

Archery was
the only event
in the very first
Stoke Mandeville
Wheelchair Games.

CREDIT: COURTESY OF
WHEELPOWER/STOKE
MANDEVILLE STADIUM

Wheelchair
basketball was one
of the first sports
trialled at Stoke
Mandeville.

CREDIT: COURTESY OF
WHEELPOWER/STOKE
MANDEVILLE STADIUM

No longer were the Games confined to military veterans. Some 400 wheelchair athletes from 23 countries competed in eight sports: more significantly, they followed the Rome Summer Games and used the same venues. This event set a template for the future and was later acknowledged as the 1st Paralympic Games.

Another milestone was the staging of the first Winter Paralympic Games in Sweden in 1976. This was also the first Paralympic Games to include athletes who were disabled but not in wheelchairs.

Over the years more sports were added and new categories of disabilities were established. Currently there are ten categories of allowable disabilities, which are then broken down into a number of classifications, depending on the sports involved.

Currently there are 22 sports on the Summer Paralympics programme and five at the Winter Paralympic Games. At the 2016 Rio Summer Paralympics 4,342 athletes from 159 countries took part. Since the Seoul Games of 1988 both the Winter and Summer Paralympic Games take place almost immediately after the respective Games, and catering for the needs of all athletes, able and disabled, is one of the key challenges for Olympic stadia designers.

One of the first, if not the first, stadium purpose-designed to fit the needs of disabled athletes was the Stoke Mandeville Stadium, developed primarily by Guttmann and still an

The Opening Ceremony at the 2016 Summer Paralympics in Rio de Janeiro.
SOURCE: CREATIVE COMMONS – CREDIT TOMAZ SILVA

international centre for disabled sport. The 1984 Summer Paralympic Games were held there, the last not to follow on from a Summer Games.

Guttmann, who had become a naturalized British citizen in 1945, was admitted to the Order of the British Empire (OBE) in 1950 in recognition of his achievements at Stoke Mandeville. He continued to campaign for greater inclusion in sport for disabled athletes, efforts that were ultimately rewarded by a knighthood in 1966.

In October 1979 Guttmann suffered a serious heart attack, ultimately dying of heart failure the following March. Many thousands of handicapped athletes around the world owe him an enormous debt of gratitude: however, it is perhaps the nickname bestowed upon him by his patients at Stoke Mandeville that says most about the true nature of this dedicated man.

He may have been criticized by many within the medical profession for what was seen as inflexibility and an autocratic manner. He was impatient with those who resisted innovation, yet, to his patients and ultimately his colleagues and staff, he was Poppa Guttmann. To the world, of course, he was universally recognized as the "Father of the Paralympic Games".

CHAPTER 5

ATHENS
GREECE, 1896

Background

It is said that Pierre de Coubertin had intended that the first modern Summer Games should be held in Paris, his home town, in 1900. However, at the first ever meeting of the International Olympic Committee in that city in 1894 it was felt that six years was too long to wait.

Another version has it that de Coubertin always saw Athens as the obvious choice, and had already sounded out the Greeks as to the possibility of staging the Games there. Either way he was determined that the inaugural Games should not be held in London, although some delegates had initially lobbied for the British capital, much to de Coubertin's annoyance.

Athens had historical credibility; it even had a potential main stadium and recent experience of staging major athletics events. The so-called Zappas Olympics – after the name of their main sponsor – had taken place there in 1859, 1870 and 1875. Although, like the ancient Games, they were national rather than international events, they did attract large crowds and were very popular.

The award of the Games was well received by the Greek public, but the government was lukewarm. Doubts about the costs involved soon surfaced. In the end Crown Prince Constantine took on the chairmanship of the organizing committee and set about raising the necessary funds. His efforts were ultimately successful. There was a particularly useful contribution from George Averoff, another extremely rich businessman.

Competitor numbers are unclear, as is the number of countries represented. There was only one national team, that of Hungary. Most of the athletes were well-to-do college students: some took part simply because they happened to be working or holidaying in Athens at the time of the Games. Officially there were 241 competitors, all male, from 14 countries, competing in nine sports. Other sources differ.[1]

All were technically amateurs, with the exception of the "fencing masters" who competed in the three fencing events. There was no Olympic Village, all the competitors had to find and fund their own accommodation.

Around 60,000 spectators attended the opening ceremony, and attendances throughout the Games were higher than ever seen at previous international sporting events of the modern era.[2]

Towards the end of the Games King George of Greece made a speech in which he proposed that the Summer Games should be held in Athens in perpetuity. He also attended the formal

At the 1896 Summer Games winners received silver medals rather than gold.
SOURCE: PUBLIC DOMAIN

closing ceremony, at which he awarded silver medals to event winners and copper medals to those placed second. Third placed competitors received nothing.

Stadium origins

Sometime around 340 BCE a natural ravine running down to the River Ilissos was remodelled to create a rectangular arena, open at one end, with the earth slopes of the other three sides terraced to provide seating for the spectators – of both sexes, contrary to earlier beliefs (see Chapter 1). This became the home of the Panathenaic Games, which until then had been held far from the city centre.

These ceased after the Roman conquest of Greece in 148 BCE. Athens became a political and economic backwater. However, during the reign of the Emperor Hadrian (AD 117–138), the city enjoyed something of an artistic and intellectual revival. This led to a major reconstruction of the original Panathenaic Stadium, funded by the enormously rich Herodes.

A horseshoe plan was adopted, with the closed end of the original stadium reshaped as a sphendone, a semicircular tier of terraces. The entrance acquired a propylon – a monumental gateway – and a three-arch bridge gave much improved access across the river.

All these were constructed of local marble, as were the new rows of seats, and there were numerous statues in marble, bronze and even gold.

Although there were no more Panathenaic Games, athletics still took place in the new stadium, along with typically Roman entertainments such as gladiatorial combats and wild animal hunts. With the arrival of the Christian era these were abandoned, and the stadium fell into disuse and disrepair. Its marble was slowly recycled, and by the early 19th century there was little left except for the foundations.

During the second half of the19th century, however, a popular movement to revive the ancient games gathered strength. Eventually the successful businessman Evangelis Zappas offered to fund the entire project, and provide cash prizes for the winners. There was opposition from many Greek politicians who feared that the revival of an ancient pagan festival would damage the image of modern Greece, and who would have preferred to showcase modern technology. Nevertheless, King Otto gave the go-ahead in 1856.

Zappas had hoped to hold the first Games in a reconstructed Panathenaic Stadium in 1859, but the work was not completed in time. Instead, the events were held in a square in Athens itself. Work continued, but Zappas never lived to see the stadium used for the 1870 Games, dying in 1865. The final Zappas Olympics were held in 1875.

Design team (for 1895 improvements)

Ernst Ziller (1837–1923) was born in Saxony but later became a Greek citizen. He designed many royal and state buildings, including the National Theatre of Greece.

Anastasios Metaxas (1862–1937) was responsible for the detailed design and supervision of construction. He was a noted shooter, and would compete in four Summer Games, winning a bronze medal for trap shooting at the 1908 London Games.

Design

Excavations in the 19th century revealed the Roman-era horseshoe shape, with a running track 333m long and 33m wide.

The first phase of reconstruction in the 1860s amounted to little more than the clearing and levelling of the site and the installation of wooden benches to accommodate 30,000 spectators, an unprecedented number for the time. During the Zappas Olympics of 1870 and 1875 the stadium was filled to overflowing on several occasions, conclusively demonstrating that it was possible to attract large crowds to athletic events, even without international competitors.

For the 1896 Summer Games the architects drew up plans for a much more ambitious upgrade: an authentic recreation of the Herodes stadium (although the propylon was not revived). Marble from the same quarry as the original was used throughout.

Reconstruction mainly involved installing rows of marble benches supported by the ancient earth berms and recreating the central arena as authentically as possible. Seating capacity was nominally 50,000, all uncovered, although many more squeezed onto the 50 rows of seats for premier events.

By current standards the central arena was remarkably small, and there was no javelin competition. The 333m running track proved to be something of a major design flaw. The curves were very tight, and competitors in events longer than 200m had to face the challenge of rounding the curves more frequently than usual. Recorded times in these events were very slow. To make matters worse, competitors were also made to run in a clockwise direction, as opposed to the otherwise universal anticlockwise orientation[3] – and there were no lane markings.

Capacity crowds filled the reconstructed Panathenaic Stadium during the 1896 Summer Games.
Note the lack of lane markings on the constricted track.
SOURCE: IOC

Essentially unchanged, the Panathenaic Stadium hosted the finish of the 2004 Olympic marathon.
SOURCE: IOC

Legacy

A decade after the 1896 Summer Games the Panathenaic Stadium hosted the 1906 Intercalated Games[4] at which several of the Olympic traditions made their first appearance.

In 1968 the FIBA European Cup Winners' Cup took place in the stadium. Home team AEK Athens defeated Slavia Prague in front of around 120,000 sitting and standing spectators, believed to be a world record, for a basketball match.

Currently it has a capacity of 45,000. The archery competitions in the 2004 Summer Games were held here, as was the finish of the marathon. At every modern Olympics, however, the Panathenaic Stadium is where the Olympic Flame is handed over to begin its journey to the next host city.

Notes

1. Mallon, Bill, and Ture Widlund (1988) T*he 1896 Olympic Games: Results for All Competitors in All Events, with Commentary* (PDF).
2. Young, David C. (1996) *The Modern Olympics: A Struggle for Revival*, Baltimore, MD: Johns Hopkins University Press.
3. https://www.sports-reference.com/olympics/summer/1896/ATH/
4. See Appendix B.

CHAPTER 6

PARIS
FRANCE, 1900

Background

Although Paris had been awarded the 1900 Summer Games at an 1894 meeting of the IOC, the success of the Athens Games almost changed the history of the Olympic Games for ever. There was heavyweight lobbying for the Greek capital to become the permanent home of the Summer Games. Even the Greek king joined in. Pierre de Coubertin came under intense pressure, but refused to yield.

Eventually the Greeks were placated by the award of the Intercalated Games on a permanent basis.[1] Then de Coubertin made a decision he later came to regret.

In order to raise the public profile of the Games, de Coubertin approached the organizers of the 1900 Exposition Universelle, or World Fair, with proposals to hold the Games alongside the high-profile event. This was agreed, but after some intense political manoeuvring de Coubertin found himself sidelined, and a new organizational committee took over from the IOC.

This scheduled a very wide range of competitions,[2] from kite flying to fire-engine racing, of which only 19 were to be eventually recognized as Olympic sports. Few were held under

There was little publicity for the 1900 Summer Games, as it was staged as part of the Paris World Fair.
SOURCE: PUBLIC DOMAIN

the Olympic banner, being billed variously as "International Championships", "World Championships" and so on. For the first and only time live pigeons were used as targets in trapshooting events: nearly 400 were killed, but the IOC eventually refused to recognize this as an Olympic event.

As a result, some athletes were for years unaware that they had competed in the Olympics. Publicity for the Games was minimal, so few spectators turned up, and organization was chaotic.

Nearly 1,000 athletes from 24 countries are believed to have taken part in the Games, although records were so poor that the true figure will never be known. This included women for the first time. Professional fencers also competed for cash prizes.[3]

There were no formal opening or closing ceremonies, and the event dragged on for five months. French athletes won most Olympic events, but a strong team of American college athletes dominated the athletics.

Origins
De Coubertin tried to convince the organizers of the World Fair to fund the construction of a near-exact replica of the ancient site of Olympia, complete with stadium, gymnasia and

There was no Olympic stadium in 1900. Events were held at a number of existing or improvised venues. The tug-of-war final was staged in the Bois de Boulogne.

temples. This was swiftly rejected as "an absurd anachronism".[4] Once the new committee had taken over it soon decided to use existing facilities scattered across Paris.

Water sports took place in the strong currents of the River Seine. The Velodrome de Vincennes housed cycling, soccer and gymnastics – and cricket and Rugby Union. Athletics, however, the heart of any Summer Games, were relegated to the Croix-Catalan Stadium, the home of the Racing Club de France soccer team.

This was a grass field, dotted with trees, rather bumpy by all accounts and frequently muddy. Competitors in track events found this very daunting after the cinder tracks they were used to. Track length was also unusual at 500m.

Hurdles were recycled telephone poles. Hammer, discus and javelin throwers often saw their best efforts stuck up a tree. Because of the lack of publicity and confusion over the scheduling, few spectators or journalists turned up – and many athletes actually missed their events.

Legacy

There was no physical legacy from the Games. The chaos and confusion did little for the Olympic movement, in fact de Coubertin later commented: "It's a miracle that it survived that

Facilities for athletics were improvised and primitive, as shown by the finishing line for the marathon, on grass.
SOURCE: PUBLIC DOMAIN

celebration." Nevertheless, the next two Summer Games, in St Louis and London, would also be held as part of a major exhibition, and it would not be until the 1912 Summer Games in Stockholm that the host city had the confidence to organize a stand-alone event.

Notes

1. See Appendix B.
2. De Wael, Herman (2003) "Olympic or not?", *Journal of Olympic History*, January.
3. Mallon, Bill (1998) *The 1900 Olympic Games: Results for All Competitors in All Events, with Commentary*. ISBN 978-0-7864-4064-1.
4. Cropper, Corry (2008) *Playing at Monarchy: Sport as Metaphor in Nineteenth-Century France*, Lincoln: University of Nebraska Press.

CHAPTER 7

ST LOUIS
USA, 1904

Background

In acknowledgement of the USA's major contribution to the Paris Olympics, the IOC decided that the 1904 Summer Games should be held in North America. New York and Philadelphia were both considered before the Games were initially awarded to Chicago.

However, the decision outraged the organizers of the 1904 St Louis World's Fair (officially known as the Louisiana Purchase Exposition). The Games would clash with their well-funded event, during which they had planned to hold various sporting competitions.

To pressurize the IOC into changing its decision the Fair organizers strove to do a deal with the American Amateur Athletics Union to hold their annual track and field championships in St Louis during the Fair. This would have meant few elite US athletes competing in Chicago, a situation the IOC could not contemplate.

President Theodore Roosevelt also favoured St Louis. Eventually, de Coubertin caved in, and St Louis secured the nomination.

Once again the Games would be relegated to nothing more than a sideshow at a World's Fair. The 94 events

As happened in Paris four years earlier, the 1904 Summer Games were overshadowed by a World's Fair.
SOURCE: PUBLIC DOMAIN

recognized as Olympic sports by the IOC were interspersed with other events, such as baseball tournaments, team and handicap races in athletics, and the so-called Anthropology days.[1]

Officially the Games lasted for four and a half months. Total participation was 645 men and 6 women (all archers) from 12 countries. This was the first time that women had appeared at a Summer Games. However, few European athletes made the arduous journey to St Louis, and only 42 of the 94 events featured foreign competitors.[2] De Coubertin himself pointedly failed to appear – as did the President.

Stadium origins

Francis Field and the adjacent Francis Gymnasium were originally commissioned by Washington University for its own long-term use following its role in the World's Fair. The University leased the site of the World's Fair to its organizers, and funded the $65,000 (approximately $1.9m in today's money) construction cost of the stadium and gymnasium out of the proceeds. Most of the recognized Olympic events took place in Francis Field, while the Gymnasium hosted the boxing and fencing competitions.

Tucked away in a far corner of the fairground and overshadowed by more popular and glamorous exhibits, the stadium epitomized the low status of the Olympics amongst the general public at that time.

Design team

Cope and Stewardson was a firm of Philadelphia architects with a reputation for academic buildings and campus designs. Washington University eventually commissioned more than ten Cope and Stewardson buildings apart from the stadium and gymnasium.

Design

A very early example of reinforced concrete construction, Francis Field was built around a "three laps to the mile" cinder running track – total lap length 536m. Its width of 6.1m eliminated the need to hold heats for the 400m and 800m: so few competitors turned up that all could be accommodated in one race.

Design was basic. Spectators sat out in the open upon concrete "bleachers",[3] with only the VIP box protected by a temporary awning. Compared to the extravaganza of the World's Fair Exhibition, with its large pavilions, artificial lakes and a Ferris wheel, much of it in a historical "Pastiche" style, there is little to say about the architecture of the stadium.

Part of one side of the arena was formed by the Francis Gymnasium, whose windows offered some of the best views, of the sprints in particular. For the first and only time a 19m x 9m roque[4] court appeared in an Olympic stadium.

Design spectator capacity was not recorded, but only 3,000 spectators attended the low key opening ceremony. Other evidence suggests that its maximum capacity was little more than 4,000.

The 1904 Olympic stadium had a maximum spectator capacity of little more than 4,000.

SOURCE: PUBLIC DOMAIN

A snowy Francis Field in 2009.

SOURCE: CREATIVE COMMONS – CREDIT SHUBINATOR

Legacy

Francis Field and Gymnasium are now US National Landmarks. The stadium has hosted American football since the Olympics and is the home of the Washington University Bears team. Temporary wooden stands increased its capacity to 19,000 until a renovation in 1984 reduced this to just 4,000. In 2004 the grass on the infield was replaced with an artificial grass surface.

Over the years the Francis Gymnasium has hosted no fewer than four presidential debates, the last in 2016.

Notes

1. Either a crude attempt to demonstrate the innate superiority of the white race or a misguided anthropological research project, the two Anthropology days featured "ethnic people" from the many "human zoos" at the Fair competing in a range of events. As most of the "competitors" were unfamiliar with Western-style athletics and had long since abandoned such pursuits as spear throwing and hunting with the bow, their performances were inadequate enough to confirm the racist beliefs of the time.
2. Mallon, Bill (1999) *The 1904 Olympic Games: Results for all Competitors in all Events with Commentary*, Jefferson, NC: McFarland.
3. Bleachers are uncovered terraces of benches, often without backrests. Traditionally constructed of wood with sun-bleached plank seating, usually open to the ground below, bleachers are now seen as low status accommodation.
4. Roque is an American development of European croquet, played on a 19m x 9m hard sand or clay court surrounded by a low boundary wall. Players are allowed to carom their balls off this wall.

CHAPTER 8

LONDON
UK, 1908

Background

When it met in London in 1904 the IOC was aware that its choice of host city for the 1908 Summer Games would be critical. Few overseas athletes had attended the St Louis Summer Games that year, damaging the Games' credibility. A much more accessible host city was required, one whose fame and reputation would add to its attraction to competitors from all over the world.

London, Berlin and Milan lost out to Rome in the bidding. By 1906, however, Italian preparations for the Games were in disarray and it seemed unlikely that Rome would be ready in time. Then Mount Vesuvius erupted, devastating large areas of the city of Naples. Funding intended for the Games had to be diverted to Naples' reconstruction, probably much to the relief of the Italian organizers. London was swiftly chosen as the only viable alternative.

The 1908 Summer Games were the longest in history, thanks to the inclusion, for the first and only time, of figure skating on ice. Four figure skating events were held on an ice rink in central London in October 1908. Overall, the Games lasted six months and four days.

In all, 37 women and 1,971 men from 22 nations took part; a massive improvement over St Louis. Women competed in the Olympic Stadium for the first time, albeit only in archery, which attracted few spectators. London was also the first Games at which the teams marched into the stadium behind their national flags[1] – and the last at which all judges and officials came from the host country.

Origins

The choice of London as the replacement for Rome was influenced by the existing plans for the Franco-British Exhibition of 1908. Held to celebrate the Entente Cordiale of 1904,[2] the exhibition took place on 57ha of farmland to the west of London. More than 100 elaborate white-painted pavilions in the Oriental style surrounded a central artificial lake; there were canals, funfair rides and a scenic railway.

Plans for what was swiftly dubbed the "Great White City" had originally included a modest athletics stadium, albeit in the prevailing Oriental style. The British Olympics Association persuaded the exhibition organizers to build a larger facility suitable for the

Summer Games at their own expense, in return for a share of the gate receipts. This was officially known as the "Great Stadium", but soon became better known as the White City Stadium.

More than 6m visitors attended the Exhibition, and the site went on to host four more similar exhibitions, before being progressively demolished during the course of the 20th century.

Design team

James Black Fulton (1875–1922) was considered to be one of the finest architectural draughtsmen of his generation. Born and educated in Scotland, he moved to London in 1894, finally setting up his own practice in 1906. One of his first major commissions was the Palace of British Applied Arts at the Franco-British Exhibition.

Structural engineer John James Webster (1845–1914) was well known as a bridge designer, responsible for one of the first reinforced concrete bridges in the UK, at Warrington. Perhaps his crowning achievement in this field was the iconic Widnes–Runcorn Transporter Bridge, the largest of the type ever built anywhere in the world.

This opened in 1905, and attracted many plaudits. Webster also designed a number of piers around the coast of the UK. His extensive experience with steel structures had a major influence on his design for the London stadium.

Originally the stadium for the 1908 Summer Games was known as the Great Stadium.
ORIGIN: PUBLIC DOMAIN

Design

Speed of construction was the absolute priority, aesthetics were of lesser importance. There was neither the time nor the funding to embellish the stadium in any way. Webster came up with a modular steel-framed structure, with 50mm thick reinforced concrete platform units spanning nearly 7m between inclined rolled steel joists. These joists were supported by latticed steel columns braced by "continuous channel bars". There were distinct echoes of seaside pleasure pier construction in the structural concept.

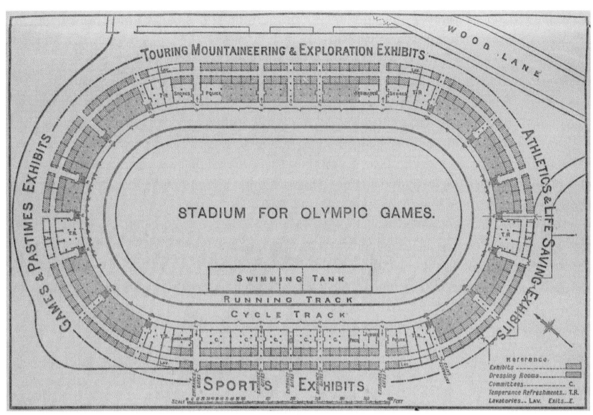

The stadium as originally planned. Note that only "temperance", i.e. non-alcoholic, refreshments were to be sold.
SOURCE: PUBLIC DOMAIN

This stadium was a great leap forward and a template for many subsequent Summer Games stadia. The water polo goals and the diving platforms in the swimming tank are clearly visible in this aerial photo, as is the sparse attendance. Part of the "Great White City" can be seen top left.
SOURCE: PUBLIC DOMAIN

At the time there were no standard dimensions for running tracks. Webster followed the precedent of St Louis, adopting a "three laps to the mile" design, i.e. a 536m lap, with a track width of 7.3m. Outside this was a banked 11m wide 600m long cycle track.

To accommodate these an unusually large footprint was needed. Final dimensions of the stadium were 305 x 181m. Within the spacious infield there was a 100m "tank" for swimming, water polo and diving events, and raised platforms for wrestling and gymnastics.

The tank was the first purpose-built Olympic pool in history, and the only one to be located in the main stadium. At previous Games swimming events took place in open water – but the new pool was unheated, unfiltered and not chlorinated, rapidly turning green and remaining chilly throughout.

Official spectator capacity was 68,000, all seated. There were complaints during the Games that the cheapest seats, at the top of the terracing, were much too far away from the action in the pool or on the platforms.

Contractor George Wimpey erected the stadium in just 10 months at a cost of £60,000 – about £3.5m in today's money. This low cost reflected the lack of such modern facilities as a public address system, floodlighting and electronic timing.

Legacy

Perhaps the most enduring legacy of the 1908 Summer Games is the official distance of the marathon, 42.195km (26 miles, 365 yards). By Royal request the 1908 event started beneath the Royal nursery windows in Windsor Castle to the west and finished in the stadium immediately in front of the Royal box. This distance was formally adopted as the international standard in 1921.[3] For all other distance events the British agreed to the use of the metric system, which became standard practice at all subsequent Games.

In 1927 the stadium was taken over by the Greyhound Racing Association. New covered terracing was built and a restaurant added. Greyhound racing was to continue until the stadium finally closed in 1985, with crowds of more than 90,000 for premier events.

Motorcycle speedway found a home there, as did many championship boxing matches. Football and rugby league teams also based themselves at White City for various periods.

Athletics was not forgotten. A 440yd (402.3m) running track was installed in 1931 and used for the Amateur Athletics Association Championships until 1970. In 1934 the stadium hosted the second British Empire Games and the fourth Women's World Games.

Notes
1. Many of the rituals introduced in London were derived from the successful 1908 Intercalated Games (see Appendix B). These included the opening and closing ceremonies, the entry of the athletes behind their national flags and the raising of the national flags of the victors.
2. Entente Cordiale is the collective name for a series of agreements signed in 1904 between the UK and France. It led to a significant improvement in Anglo-French relations.
3. Jolly, Rhonda (2008) "The modern Olympics: an overview", 3 June. Research paper, Parliament of Australia, Parliamentary Library (PDF).

Italy's Dorando Pietri crossed the finishing line of the marathon first, but was disqualified for receiving assistance
to finish. The length of all subsequent marathon races was established at these Summer Games.
SOURCE: PUBLIC DOMAIN

CHAPTER 9

STOCKHOLM
SWEDEN, 1912

Background

When they met in 1909 to select the next host city for the Summer Games, IOC delegates were anxious to avoid a repeat of the 1906 Rome debacle. Although the late switch to London had saved the 1908 Games from ignominious cancellation, no such fortuitous alternative was likely to present itself in 1912.

Staging the Games as just another sideshow event at a World's Fair or similar event, as in Paris, St Louis and London, was now seen as a counterproductive option. De Coubertin, perhaps with the spectres of Paris and St Louis in mind, called for future Games to be: "more purely athletic, more dignified, more discreet, more in accordance with classic and artistic requirements, more intimate, and, above all, less expensive."

In other words, perhaps, no more marble stadia, as at Athens 1896, no more ballooning or fire-engine racing, as at Paris, no more of the Anthropology days that tarnished the St Louis Games. What the IOC needed above all was a host city that could guarantee it had the financial resources to stage a Games that would meet all the Olympic objectives.

Sweden had expressed interest in staging the Games immediately after London 1908. Two Swedish members of the IOC soon persuaded major sports governing bodies to join them in petitioning the government to fund the estimated 415,000kr (approximately £2.6m in today's money) cost of hosting the Games.

This support was forthcoming. At the 1909 IOC meeting the Swedish delegates put forward the only realistic proposal. Berlin was also interested, but the German IOC representative accepted an assurance that Berlin would get the 1916 Games.

In all, some 2,408 competitors from 28 nations took part in Stockholm, including a team from Japan, the first time an Asian country had been represented. There were 48 female competitors, and women's swimming and diving made their first appearance. More than 100 events in 14 sports were held: boxing was not included, as the Swedish authorities considered it a barbaric sport.

In a nod to de Coubertin, art competitions were introduced into the Olympic schedule. There were awards for literature, sculpture, architecture, music and painting: Pierre de Coubertin, who entered under a nom de plume, received the gold medal for literature.

More lasting innovations were the introduction of electronic timing and a public address system. It was also the last Games at which the 'gold' medals were solid gold.

It was noted that these Summer Games were a more harmonious event than London 1908, with the bitter rivalry between the United States and Great Britain teams far less evident.[1] This may be down to the appointment of international judges and referees for the first time.

Stadium origins

Initial plans for a new stadium were relatively modest. A temporary timber structure was envisaged, located on the existing site of the Stockholm Athletic Grounds, with the estimated 400,00Kr cost to be met by a national lottery.

The chosen site was centrally situated with good tramway communications, and several existing sports facilities nearby.

Perceived fire risks, however, cast doubt on the timber design, and architect Torben Grut was asked to develop a stone alternative. This turned out to be prohibitively expensive, so a much cheaper mixed brick, concrete and stone solution was finally adopted.

Design team

Torben Grut (1871–1945) was an established architect and also a well-known sportsman. He was Swedish tennis champion in 1896, and his son William won the modern pentathlon prize at the 1948 Olympics.

The stadium during the 1912 Summer Games, with an upper tier of temporary seating at the northern end.
SOURCE: PUBLIC DOMAIN

Design

Although conveniently located, the site was only just large enough for the projected stadium. There was a rocky outcrop at the northern end, and while rock underlaid much of the site at a maximum depth of 11m, the subsoil was mostly "boggy" sands and clays. Original plans to found the entire structure on the rock were abandoned due to time and cost constraints: instead a combination of concrete columns founded on the rock and a raft foundation was adopted.

At the northern end the rocky outcrop was incorporated into a series of terraces, topped by a second temporary tier. Basements proved a particular challenge, given the waterlogged subsoil.

Overall, the design paid some homage to the original Classical Greek model, with a U-shaped arena closed by an arcade to the south. The cinder running track, however, took a more modern symmetrical form. In the Official Report the track is described as having a "circumference" of 384.10m: no further information is given.

One seminal innovation was the construction of tunnels that allowed competitors and officials to reach the central arena without crossing the running track.

Most of the spectator seating was open air, but the side spectator areas were partially covered. There were two towers, one each side of the stadium, offset to the north. One was a clock tower, an essential feature of the public buildings of this time. Spectator capacity with the

Scandinavian expertise shows in the brickwork on the southern gateway.
SOURCE: CREATIVE COMMONS: CREDIT DERBETH

temporary seating installed was around 22,000, making it one of the smallest stadia to be used for athletics at a Summer Games.

The southern façade had an archway built in the character of a military castle, forming a gateway to the stadium. Brick was the main construction material, with the high standard of the brickwork confirming that the use of this material in Scandinavian hands had reached a fine level of maturity.

Particularly impressive were the vaulted arcades, especially at the southern end. The completed stadium showed an architectural character and confidence unusual in its time. Bearing in mind that London 1908 broke new ground with the first purpose-designed modern Olympic stadium, the Stockholm stadium just four years later was a remarkably well-composed and mature design. Final cost was around £15m in today's money.

Later, Pierre de Coubertin commented: "The Gothic Stadium with its pointed arches and its turrets, its technical perfection, its good order and its purposeful disposition seemed to be a model of its kind."[2]

Legacy

Overall, the Stockholm Games inspired a fresh start in Olympic architecture, and also, through the art competitions, re-established a link between sport and the arts. They can be regarded as the close of the first phase of the development of the modern Summer Games, a phase terminated by the First World War.

The stadium today, with the northern tier of temporary seating removed.
SOURCE: CREATIVE COMMONS: CREDIT JOHANNES SCHERMAN

Soon after the Games were over the upper level of the north-east stand was removed, reducing capacity to around 19,000 for sporting events, and 30,000 for concerts. Otherwise the original design is virtually unchanged, except for the installation of a modern 400m running track surface, although official capacity is now only 14,500.

Over the years the stadium has hosted many international athletic events, and more world records were set there than anywhere else in the world, until the Beijing Summer Games 2008.

Quarantine restrictions made it impossible to hold equestrian events at the 1956 Melbourne Games. The Stockholm stadium acted as the substitute venue.

Notes

1. Keating, Frank (2012) "Stockholm 1912 set the gold standard for the modern Olympics", *The Guardian*, 1 May.
2. Pierre de Coubertin (2000), *Olympism: Selected Writings*, edited by Norbert Müller, International Olympics Committee, Lausanne.

CHAPTER 10

ANTWERP
BELGIUM, 1920

Background

Belgium first expressed an interest in staging the Summer Games at the March 1912 meeting of the IOC in Switzerland. No particular city was put forward. It was not until 1914 that Antwerp became the official bidder in competition with Amsterdam, Budapest and Rome, but no final decision was taken before the First World War broke out.

The 1916 Games had been awarded to Berlin in 1912, but never took place. Antwerp was occupied by the Germans from 1914 to 1918 and its citizens endured severe privations. After the Armistice there was a widespread conviction that Antwerp had a moral right to the 1920 Games, a view strongly supported by France.

At the 1919 meeting of the IOC there were originally no fewer than eight bidders, despite the short notice. Antwerp was again challenged by Amsterdam and Budapest: fresh into the fray were Atlanta, Cleveland and Philadelphia, and Havana. Lyon dropped out before the final vote.

Publicity for the 1920 Summer Games was limited by post-war paper shortages: it was claimed this was a significant factor in the disappointing attendances at many events.
SOURCE: PUBLIC DOMAIN

Announcing the award of the Games to Antwerp, the IOC stated it was "a compensation and to honour the Belgians who fought, suffered and died during the war."

From this point onwards "known historical facts are few, incomplete and often inaccurate".[1] This is down to a financial meltdown in the organizing committee immediately after the Games and the consequent failure to publish a comprehensive Official Report within a reasonable timescale.

"Some years later" – to quote from the IOC archives – a quasi-Official Report in the form of 178 stencilled pages in a ring binder was finally published. Available only in French, it gave few details of the stadium and associated infrastructure.

However gratifying the award of the Games might have been, in practice it gave the local organizing committee little more than 12 months to organize and stage the Games. Preparations were hindered by, amongst other factors, the flood of returning refugees and a shortage of paper that curtailed advertising efforts. Providing accommodation for competitors and their coaches was a logistical nightmare. Several ships in the harbour were pressed into service to provide basic facilities.

Banned from the Games were the recent enemy states of Germany, Austria, Hungary, Bulgaria and the Ottoman Empire (now Turkey). Nevertheless, 2,561 men and 65 women turned up to compete in 22 sports, including ice-skating and ice hockey. After refusing to send a women's swimming team to Stockholm on the grounds it would be "obscene", the United States sent a six-woman team to Antwerp, which won two gold medals.

Innovations at the Opening Ceremony included the first voicing of the Olympic Oath, the first flying of the Olympic flag and the first ever release of doves.

Incessant bad weather, high ticket prices and very limited advance publicity hit attendances in the main stadium. The only time it was full was for the football final between Belgium and Czechoslovakia, when local youths actually tunnelled into the stadium, allowing thousands to watch for free. Belgium won by default when the Czech team marched off the pitch following the sending-off of one of its defenders.

After the Games there was much local press criticism of the poor publicity and communications and the disappointing spectator attendance at what were seen as aristocratic events, such as athletics, sailing and show jumping. Only boxing and football attracted significant audiences.

In response the organizers pointed to the limited coverage given to the Games in the local, national and international press. With the benefit of hindsight, however, the Antwerp Summer Games can be seen as crucial to re-establishment of the Olympic tradition after the social and political disruption of the First World War.

Stadium origins

Initial plans for the main stadium located it on an existing "exercise field". However, the powerful and aristocratic Beerschot Athletic Club lobbied hard for their existing stadium nearby to be modernized and extended, which coincidentally would involve the upgrading of road links to the area and the installation of mains water, gas and electricity supplies.

Land around the stadium would rocket in value: land that by coincidence was mostly owned by Beerschot club members. Nevertheless, so narrow was the time window that the Beerschot option was reluctantly adopted.

Design team

Architects Fernand de Montigny and L. Somers were responsible for the upgrade of the existing Beerschot stadium.

The London-based contractor Humphreys & Co undertook the construction.

Suggestions that famous British football stadium architect Archibald Leitch was involved in the design process appear to be ill-founded.

Design

Few, if any, major sports stadia have featured mature trees growing on the terraces – but what was renamed the Olympisch Stadion did. Details are scarce, but it seems likely the original Beerschot stadium, which dated back to 1900, had grassed terraces where the trees would be more at home, and some if not all of the original trees were retained. During the football final many younger fans swarmed up the trees to obtain a better viewpoint.

Mature trees growing on the terraces were a unique feature of the 1920 Olympic Stadium.
SOURCE: PUBLIC DOMAIN

The lack of a comprehensive Official Report means that information on the 1920 stadium is hard to find.
SOURCE: PUBLIC DOMAIN

Concreting over these terraces was a significant part of the Olympic upgrade. A second grandstand was constructed opposite the existing stand, which was extended to the full length of the straight. The structure of the grandstands was basic, with simple roofs fronted by columns in front of the spectators.

This has echoes of the style of Archibald Leitch, which may account for some sources attributing the stadium design to him.[2] However, they lack the distinctive Leitch pediment, and the designers of the Olympic upgrade seem to have simply followed the design of the original stand.

Terraces were backed with colonnades in the neo-classical style: this is unsurprising given the aesthetic development of the Olympic Movement thus far.

Sources differ markedly on the Olympic spectator capacity. A 1919 upgrade of the Beerschot Stadium is said to have created a 21,000 capacity venue. The same source implies that the Olympic upgrade increased spectator seats from 3,000 to 10,000, for a total capacity of 20,000. Another source gives a figure of 27,250,[3] while a third insists on a curiously precise figure of 12,771.[4]

An "English specialist" called Perry is said to have built the cinder running track. He is also credited with having built the track at the 1912 Stockholm Summer Games, where he is described in the Official Report as a "Charles Perry of Stamford Bridge" (see Chapter 9). There are varying reports on how badly it was affected by the incessant rain.

Also unclear is how much of the original stadium's structure was retained. It appears that "numerous other 'decorative' features such as the main entrance and the marathon entry were temporary constructions made of plaster."[5] These were neo-classical style triumphal gates,

These impressive structures were said to be only lath and plaster.
SOURCE: PUBLIC DOMAIN

which do appear to have a stucco exterior: given the timescale of the project it seems unlikely that monumental stone structures could have been possible.

Legacy

After the Games it soon emerged that the financial underpinnings to the event were simultaneously obscure and fragile. The Beerschot Club received back their much-improved stadium almost for free as compensation for allowing it to be used for the Games. The organizers were reduced to begging the government for extra funding: this was refused, and liquidation was the only possibility.

A stadium still stands on the site of the 1920 original, but it has undergone massive changes over the decades. The home since 1921 of the Beerschot AC football club, and still known as the Olympisch Stadion, the remodelled venue now has stands on all four sides and no running track. Spectator capacity is less than 13,000, all seated.

Most of the changes took place in 2000, following Beerschot's merger with another local club. There had been earlier partial demolitions since the Games.

Notes

1. With no contemporaneous Official Report available, the main source for this chapter is *Antwerp 1920 – The Games Reborn*, by Roland Renson, published by Pandora in 1996.
2. http://olympics.ballparks.com/1920Antwerp/index.htm
3. Inglis, Simon (1990) *The Football Grounds of Europe*, London: Willow Books.
4. *Sport-Revue*, 10 February 1920.
5. Ibid.

CHAPTER 11

PARIS
FRANCE, 1924

Background

This was to be the last Summer Games under Pierre de Coubertin's presidency of the IOC. After the chaos and confusion of Paris 1900, de Coubertin was anxious to wipe the slate clean and leave on a high note. It was therefore no real surprise when, in 1921, Paris beat off challenges from Amsterdam, Barcelona, Los Angeles, Prague and Rome to become the first city to host the Olympics twice.

Although Germany was still shunned for its aggression during the First World War, there were 44 nations represented at the Games, including the first appearance of Ireland as an independent nation. A total of 3,089 competitors, including 135 women, took part in 126 events in 23 disciplines. For the first time ever there was an "Olympic Village" in the shape of temporary wooden huts close to the stadium.

This was the first Games at which a 50m pool with marked lanes was used for the swimming events.

Athletes were offered three meals a day – although the British team insisted on bringing its own chef – and there was a foreign exchange, shops, a post office and even a hairdresser. The American team apparently rejected the concept of communal accommodation and set up its own camp nearby.

Significantly, around 1,000 journalists from all around the world attended these Games, a mark of their increasing popularity and public profile.

Highest attendance of the Games was the claimed 40–50,000 who watched the Rugby Union final between France and the underdogs of the USA. The French were heavily defeated, their fans rioted, and attacked and severely injured American spectators and reserve players. Fifteen-man Rugby Union was immediately thrown out of the Games and only reappeared in its seven-a-side format in 2016.[1]

Origins

Situated in the north western suburb of Colombes, the stadium site had been hosting sporting events since 1883. Originally it was a "hippodrome", a horse-racing venue, and then in 1907 it was bought by the daily Paris newspaper *Le Matin*.

A fairly basic new stadium with a capacity of 20,000 spectators was built and dubbed the Stade du Matin. It hosted rugby, athletics and soccer. In 1920 the stadium became the home of Racing Club de France, one of the oldest sports clubs in France. Racing Club's previous stadium, Croix-Catelan, had been the unsatisfactory athletics venue at the 1900 Games.

The architect

Louis Faure-Dujarric (1875–1943) was an established Parisian architect who also happened to be a former Racing Club player.

Design

It seems the original Stade du Matin was constructed by the long-established technique of excavating nearly 3m into the underlying clay and using the spoil to form terraced berms around the infield. At the start of the design process for the Summer Games it was thought that spectator capacity would need to be increased to 100,000; after reconsideration a more realistic target of 60,000 was adopted. Even this proved too optimistic: final official capacity during the Games was 45,000.

Layout was somewhat similar to London 1908 (see Chapter 8) although at 500m the running track was shorter and there was no cycle track.

Stade de Columbes during the 1924 Olympic football final. From the air, the similarities to the London 1908 stadium are obvious.
SOURCE: PUBLIC DOMAIN

The original Stade du Matin was upgraded with concrete terraces and a steel-framed roof.
SOURCE: PUBLIC DOMAIN

A reinforced concrete superstructure was erected on top of the existing berms. Spectators were accommodated on wooden benches. The two 144m long steel-framed grandstands on either side provided shelter to 20,000 privileged spectators, while the less fortunate had to make do with the open terraces at either end.

Following the innovation at Stockholm, there was an athletes' tunnel that led from the 40 changing rooms below the stands to the infield. The 8.5m wide track was surfaced with red cinders, carefully selected and graded to ensure stability on the clay subsoil.

The poor performance of cinder tracks is obvious here.
SOURCE: PUBLIC DOMAIN

The red track, green infield and yellow-painted grandstands combined to produce an aesthetic widely described as "gay" at the time, the term being used in its original sense. Overall, however, the new stadium received few architectural plaudits.

In 1928 it was officially renamed the Stade Olympique Yves-du-Manoir in honour of a famous French rugby player, but is universally known as the Stade de Colombes.

Legacy

For many years the stadium was at the centre of French sport. In the 1930s it was renovated and enlarged to hold more than 60,000 spectators, and hosted the football World Cup and the European athletic championships.

Post-Second World War, however, maintenance was skimped and the stadium went into decline. In 1972 a competing arena, the modern and stylish Parc des Princes opened in the south-west of Paris. International football and rugby matches and the finals of the French football cup decamped to Parc des Princes and never returned.

The stadium in 2009.
SOURCE: CREATIVE COMMONS: CREDIT DR. CLINT BEANS

The Colombes stadium continued to decay and much of it was demolished towards the end of the last century. Despite several attempts to revive its ancient glories, it still struggles to find a 21st-century role, and its future remains uncertain.[2] Currently there is a proposal for it to host the field hockey events at the 2024 Summer Games.

Notes
1. http://www.rugbyfootballhistory.com/olympics.htm#usa1924
2. http://olympics.ballparks.com/1924Paris/index.htm

CHAPTER 12

AMSTERDAM
THE NETHERLANDS, 1928

Background

Amsterdam had bid for what was now known as the Summer Games three times before it was awarded the 1928 Olympics. At the 1921 IOC meeting in Lausanne it was agreed that Paris would stage the 1924 Games while Amsterdam was selected for the subsequent event. Los Angeles protested, but eventually was placated by the award of the 1932 Games.

The city's success was welcomed by most of the population, even though there was little popular support for such sports as athletics and swimming. Both the national and regional governments promised financial subsidies, and a national lottery was mooted. However, there was still a strong fundamentalist Christian Calvinistic movement in the Netherlands, which was outraged by the idea of staging what they regarded as a "heathen" spectacle in the nation's capital.

"Passions would be stirred in a very dubious way," it was claimed: "Women will lose their feelings of modesty and virtue." The government tried to defuse the situation by promising there would be no "desecration of the Sabbath" and by refusing permission for a lottery, but the Calvinists were unimpressed.

They mustered enough political support to force the government to withdraw its promise of subsidies. Regional governments followed suit. The Amsterdam Olympics Committee found itself in a nightmare situation.

To save the Games and avoid national humiliation the Committee effectively invented crowdfunding. It appealed to the nation: the response was spectacular. Every club and organization vaguely connected to sport made contributions, as did many others such as restaurant chains, chocolate makers and the post office, which issued a wide range of Olympic stamps.

Individual contributions poured in, with a significant tranche from the Dutch East Indies, and even Dutch expatriates responded. The total collected was more than enough to stage the Games.[1]

In all, 46 nations sent 277 women and 2,606 men to compete in 14 sports. Germany returned to the Summer Games for the first time since the First World War, fielding the largest team and coming second in the medal table after the USA.

Women's athletics and gymnastics were included for the first time, despite strong resistance from within the IOC and outside it. Only five track and field events were scheduled. As a protest against this minimal inclusion, the British women's team boycotted the Games.

After competitors in the women's 800m were photographed sprawled on the grass after the race it was claimed they had collapsed due to overexertion. This was used as an excuse to remove the women's 800m from the programme until Rome 1960.

Other innovations included the first appearance of the Olympic Flame; the first time the Greek team led the competitors into the stadium for the Opening Ceremony with the host nation bringing up the rear, and the first time the events were held to a 16-day schedule, now standard.

There was no Olympic Village. Competitors were housed in local hotels and boarding houses, and even on ships in the harbour.

The first-round football match between the home team and the eventual gold medallists Uruguay attracted the largest crowd of the Games. Uruguay won.

Stadium origins

At the initial planning stage it was hoped that an existing stadium to the south of the city centre could be renovated or extended to become the main Olympic stadium. However, it soon became obvious that it was too small to accommodate a full-size football pitch and both a 400m running track and a 500m cycle track in the style of London 1908 (see Chapter 8).

One alternative was to locate the running track outside the stadium proper. Another was a temporary timber design. Eventually it was realized that access to the existing stadium was so restricted that there could be serious problems during any major crowd movements.

A completely new stadium was needed. The chosen site close to the existing stadium was marshland, described in the Official Report as "not much more than a quagmire." Some 750,000m³ of dredged sand was needed to form a raised platform on which the new stadium could be constructed.

Design team

Jan Wils (1891–1972) was a well-respected Dutch architect known for his use of patterned brickwork, primary colours and clean lines. He was also an associate of Colonel Scharroo, a key member of the organizing committee, and an acquaintance of Pierre de Coubertin.

In the Official Report, design of the iron and steel grandstand roof structures is attributed to an E.A. Van Genderin Stort.

Design

Wils was the co-author of a book on sports venue design that was one of the first to stress the importance of considering the impact such facilities would have on the urban landscape that surrounded them. The Olympic stadium was designed to blend in with the surrounding architecture and to be easily accessible from Amsterdam.

Layout was "standard", following the precedent of earlier stadia, particularly London's White City Stadium, home of the 1908 Games. The running track was 400m long for the first time at a Summer Games: this was to become the standard for all future Olympics, and, indeed, eventually, for all athletics stadia worldwide. It was surrounded by a 9m wide 500m cycle track. Two tunnels gave access to the infield.

Grandstands alongside the straights had a claimed spectator capacity of up to 21,337, while originally the terraces were designed to hold more than 25,000 standing spectators. Seats were basic: wooden planks on metal brackets embedded in concrete. In practice the true capacity was around 32,000.

Spectator staircases were designed to be capable of evacuating the stadium in less than 12 minutes.

The influence of the London 1908 stadium is still clear in the classic layout of the Amsterdam Olympic Stadium.
SOURCE: PUBLIC DOMAIN

Nearly 4,500 driven concrete piles up to 18m long supported the stadium structure, which was largely constructed of reinforced concrete. The cinder, clay dust, "red soot" and ground brick running track was based on "a very heavy foundation of stone blocks" to minimize the risk of subsidence.

In the interests of safety the structure of the cycle track was independent of the main stadium structure. The gap between the balustrade on the outer edge of the track and that in front of the spectators facilitated drainage of the spectator areas: it also reduced the risk of "hats, walking sticks, coats, cigar ends, etc." falling onto the track.

More than 2 million bricks were used to form a curtain wall around the concrete structure. Where concrete was exposed it was sandblasted.

Providing a vertical accent to the largely horizontal lines of the stadium and the surrounding buildings was the concrete-framed brick-clad Marathon Tower to the east of the stadium. It supported not just public address speakers but the metal bowl where the very first Olympic Flame was lit without ceremony at the start of the Games. Locals affectionately dubbed the bowl the "KLM pilot's ashtray".

Once opened, the stadium was lauded as an outstanding example of what was known as the Amsterdamse School of architecture, combining superb functionality with aesthetic simplicity and a quiet appeal. In recognition of his design's success, Wils was awarded the gold medal for architecture by the IOC during the Games.

Legacy

Apart from a much-loved and respected stadium, perhaps the most enduring legacy of the Amsterdam Games is the near-universal adoption of the sign for car parking, the white P on a

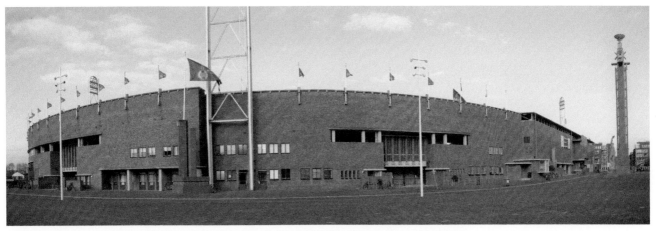

Declared a national monument in 1996, the original 1928 stadium is little changed. The Marathon Tower still supports what was dubbed "the KLM pilot's ashtray" where the first ever Olympic Flame was lit.
SOURCE: PUBLIC DOMAIN

blue background. This was created specifically for the Games, as the organizers feared parking chaos in the area around the stadium.

Upper tiers of seating were added at the northern and southern ends of the stadium in 1937. The venue hosted many sports, including motorcycle speedway.

In 1996 there were proposals to demolish the stadium: these were thwarted when it was declared a national monument. The 1937 extensions were removed, as was the cycle track, which created access for new offices below the seating tiers.

The stadium reopened in 2000. It was selected to host the 2016 European Athletics Championships, and remains a popular tourist destination.

This now universal parking symbol was designed originally for the 1928 Summer Games.
SOURCE: PUBLIC DOMAIN

Note

1. Paauw, R. and J. Visser (2008) *A Model for the Future: Amsterdam, Olympic Games 1928*, De Buitenspelers.

CHAPTER 13

LOS ANGELES
USA, 1932

Background

After losing to Paris in 1924 and Amsterdam in 1928, it was third time lucky for Los Angeles when it was awarded the 1932 Summer Games. It was the only bidder.

Los Angeles, however, already had a potential main stadium, the Memorial Coliseum, completed in 1923. It could easily be extended in plenty of time for the Games, and was well located. Other sporting facilities were also available in the area. Only one major new venue was needed, an aquatics stadium with a 10,000-spectator capacity.

Although there had been an "Olympic Village" of sorts at the 1924 Paris Games, the Olympic Village built for the 1932 Games is generally acknowledged to be the model for all subsequent Games. Located in the Baldwin Hills area, the complex of several hundred buildings included a hospital, a fire department, a bank and a post and telegraph office.[1]

Women, of course, were segregated. The Village was an exclusively male preserve; female competitors were housed in a city centre hotel.

In the event there were many fewer athletes competing than at Amsterdam. The Great Depression was raging: few countries could afford to send large national teams all the way to the West Coast of America. Just 126 women and 1,206 men turned up, representing 37 countries to compete in 14 sports. This contrasted unfavourably with the nearly 2,900 athletes from 46 countries who participated in the 1928 Summer Games.

Even President Herbert Hoover failed to appear, following the example set by President Theodore Roosevelt, who had pointedly ignored the 1904 Games in St Louis. Nevertheless, local press reports suggest that the 1932 Games made an overall profit of $1m.[2]

Stadium origins

Situated in Exposition Park in the south of the city and just across the road from the University of Southern California (USC), the Los Angeles Memorial Coliseum was commissioned in 1921 as a tribute to local veterans of the First World War. It was completed in little more than 16 months at a construction cost of just under $1m.

Exposition Park itself was known as the Agricultural Park until 1910. Its 65ha played host to farmers' markets and horse, dog and even camel racing. It was then upgraded, the racetrack abandoned, and museums and gardens replaced the unspecified "other unsavoury activities".[3]

Aerial view of the Opening Ceremony of the 1932 Summer Olympics.
SOURCE: IOC

The new stadium was to be constructed on the site of the abandoned race track. Ground conditions were free-draining sand and gravel, simplifying construction and economizing on foundations.

As first built the Coliseum had a capacity of more than 75,000, making it by far the largest stadium in Los Angeles. It immediately became home to the American football teams of the USC and the University of California, Los Angeles.

Design team

The father and son team of British-born John B. Parkinson (1861–1935) and Donald D. Parkinson (1895–1945) were also responsible for several of the city's other landmark buildings, including the City Hall and the Union Station. This despite John Parkinson having never had any formal education or architectural training, starting his career in Los Angeles as a sawmill foreman.

LOS ANGELES COLISEUM (OLYMPIC STADIUM). LOS ANGELES. CALIFORNIA 63784 T112

An Olympic postcard from 1932.
SOURCE: CREATIVE COMMONS – CREDIT TICHNOR BROTHERS

Design

Conceptually the new stadium's design paid homage to both the Ancient Greek and Roman traditions, with more in common with the recreated Panathenaic Stadium in Athens than most of its immediate predecessors. Comparisons with the Coliseum in Rome were inevitable.

A "Roman bowl" was excavated to a depth of 9.75m and the excavated spoil used to construct berms around the site – effectively "cut and fill" construction. The footprint was 317 x 225m, larger than Rome's Coliseum, and a source of some pride to local residents.

Again following the Classic model, the "bleacher"-type terraces of bench seating were supported by the sides of the excavation and the berms above. There was no covered seating; the entire venue was open to the elements, but thanks to the normally benevolent local climate this was rarely a problem.

In situ reinforced concrete was the structural material of choice. At the eastern end of the stadium it was used to construct an iconic peristyle, with a central propylaeum, or triumphal arch, flanked by seven arches on each side.

For the 1932 Games a central torch structure was added above the propylaeum to house the Olympic Flame. An additional tier of seating supported by an in situ concrete frame was constructed, along with a second level of access tunnels. Around the rest of the exterior above

The propylaeum, or triumphal arch, at the eastern end of the stadium.
SOURCE: IOC

the level of the berm there was a continuous flow of pierced in situ concrete panels and pilasters.

In all there were 79 rows of seats, bringing official spectator capacity up to 101,574.

Another innovation that became a standard feature of all future Olympic Games was the victory podium. The running track was shortened from 440yds to 400m, and a much larger press box installed.

Legacy

The Coliseum made history in 1984 when it became the first stadium ever to host the Summer Games twice. By this time the original stadium had undergone a major renovation, when the original bench "bleachers"[4] were ripped out and replaced with individual theatre-style seats. This reduced capacity to 93,000.

Throughout its life the Coliseum's main function was to host American football and baseball games, for which it proved to be not well suited. Its main drawback was its size. As a baseball venue the Coliseum's infield was too small and the wrong shape, while football spectators could be a very long way from the action.[5]

In the early 1990s a radical remodelling took place. The infield was lowered by 3.4m and 14 new rows of seats replaced the running track. There were plans for an even more radical transformation – then the 1994 Northridge Earthquake caused serious damage to the stadium.

Nearly $100m had to be spent on repairs, and the radical plans were abandoned. In 2015 the USC unveiled a $270m plan to renovate and restore the Coliseum and create new amenities for spectators.

Following the award of the 2028 Summer Games to Los Angeles it was announced that the Opening and Closing Ceremonies would be held in the Los Angeles Stadium at Hollywood Park, due to open in 2020. The soccer competition would also be held there. Athletics, however, would return to the Coliseum for a record third time, which will require the installation of a new running track.

Notes

1. www.baldwinhills.info/olympicvillage.php
2. Zarnowski, C. Frank (1992) "A look at Olympic costs", *Citius, Altius, Fortius,* summer (PDF).
3. Hobbs, Charles P. (2014) *Hidden History of Transportation in Los Angeles,* Charleston, SC: The History Press. ISBN 978-1-62619-671-1.
4. Bleachers are uncovered terraces of seats, often without backrests. Traditionally constructed of wood with sun-bleached plank seating, usually open to the ground below, bleachers are now seen as low status accommodation.
5. Schwarz, Alan (2008) "201 feet to left, 440 feet to right: Dodgers play the Coliseum", *The New York Times,* 26 March.

CHAPTER 14

BERLIN
GERMANY, 1936

Background

When the IOC met in Barcelona in 1931 there were no fewer than 13 cities bidding to host the 1936 Summer Games, including four from Germany: Berlin, Nuremburg, Cologne and Frankfurt. Germany had been excluded from the Olympics after the First World War: the ban was only lifted in 1925. Six years later, these German bidders were ranged against Alexandria, Barcelona, Budapest, Buenos Aires, Dublin, Helsinki, Lausanne, Rio de Janeiro and Rome.

In the event only Berlin and Barcelona received any votes, and Berlin triumphed in the first round by 43 votes to 16. This was the last time the IOC met to vote on the issue in a city that was bidding for the Games.

At the time of the award Germany was still a republic. However, the so-called Weimar republic, set up in 1919 after First World War, was overthrown in 1933 by Adolf Hitler's National Socialist Party (contemptuously dubbed "Nazi" by its opponents).

Hitler's virulent anti-Semitism and militaristic stance caused major disquiet amongst IOC members. When it was announced that the persecution and demonization of German Jews and Roma (gypsies) would extend to banning them from representing "Aryan" Germany at the Olympics, there was serious consideration given to taking the Games away from Berlin.

At the same time there were many in the National Socialist Party who condemned the Olympics as "a Jewish international enterprise", and there was a burgeoning campaign against the Games.[1] Hitler, however, once he had achieved supreme power, saw the Games as a potential showcase for his new Germany, the Third Reich, and was determined they should go ahead.

Pressure was put on the IOC by a major US campaign for a boycott of the Games if Jewish participation was not guaranteed. In the event, after prolonged negotiations, the IOC signalled it would allow the Berlin Games to take place if there was at least one Jewish athlete on the German team.[2]

Germany's response was to select the half-Jewish Helene Mayer, who had won a fencing gold medal for Germany at the 1928 Amsterdam Summer Games. Ironically, her family had fled to America following the persecution of her Jewish physician father. Nevertheless, Mayer returned to Germany to compete, winning a silver medal in the foil competition. Tall, blond and fashionably "Aryan", Mayer's ethnicity was never mentioned in the Nazi-controlled German media.

Before the Games all Roma living in Berlin were rounded up and transported to a concentration camp. Anti-Semitic signage was removed from most tourist attractions to further "clean-up" the city.[3]

After the Nazi takeover their own newspaper had campaigned for black athletes to be banned from the Games:[4] this made the triumphs of Jesse Owens and other black American athletes hard to swallow. It seems, however, that black athletes were well received by the German public and loudly applauded by spectators. Although Hitler was reported to have refused to shake Owens' hand, in fact he had stormed out during the first day of the Games after Olympic officials insisted he must shake the hands of all gold medal winners, not just the German ones.[5]

In all, 49 nations sent 331 women and 3,963 men to compete in 19 sports. Afghanistan, Bermuda, Bolivia, Costa Rica and Lichtenstein appeared for the first time. Handball and basketball made their Olympic debut.

Although Olympic Flames had been lit at Amsterdam in 1928 and Los Angeles in 1932, Berlin was the first city to stage an Olympic Torch relay from Greece. Intended to showcase the efficiency and organizational skills of the Third Reich, this first relay involved more than 3,000 runners over more than 3,000km in 12 days.

These Games were the first to be shown live on television, although coverage was limited.

Stadium origins

Berlin had been awarded the 1916 Summer Games during the 1912 Stockholm Olympics, only for the First World War to intervene. As part of the preparation for these Games work began on a new stadium in Charlottenburg, to the west of Berlin, within an existing horse-racing course. A new suburban rail station opened nearby in 1913.

Ground conditions – free-draining sand – encouraged the adoption of a classic design with the central arena sunk below ground level. Overall, what would be known as the Deutsches Stadion followed London's 1908 White City Stadium model (see Chapter 8). There was a 600m running track inside a 665m velodrome, but no central swimming pool, and the architecture was rather more monumental than in London.[6]

Official capacity was 33,000 spectators, although more than 60,000 squeezed in for a football final in 1923.

Originally the German government had planned simply to refurbish and upgrade the Deutsches Stadion to act as the main stadium for the 1936 Games. When Hitler came to power, however, he demanded something much more impressive. The old stadium was demolished in 1934 to make way for a massive new Olympics complex.

Design team

Architect Werner March (1884–1976) was the son of Otto March, the designer of the original Deutsches Stadion. He was assisted by his brother Walter.

Hitler is said to have seen March's original design as too Modernist, and charged interior minister Wilhelm Frick with the responsibility of ensuring that something much more classical and Aryan was produced. Frick in turn delegated the task to architect Albert Speer (1905–1981), later to become Germany's wartime Minister of Armaments.[7]

There is no mention of a structural engineer in the Official Report.

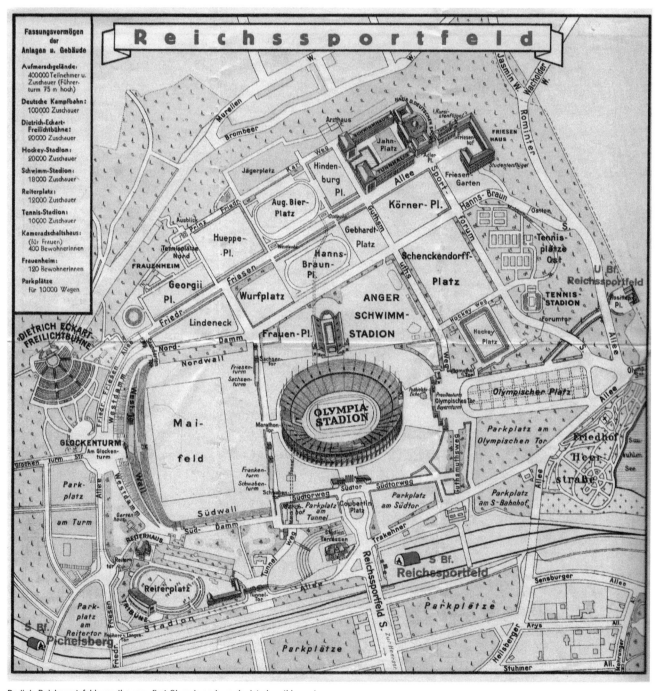

Berlin's Reichsportsfeld was the very first Olympic park, as depicted on this contemporary map.
SOURCE: PUBLIC DOMAIN

Design

For the first time ever, the main stadium would be part of an integrated sports complex, dubbed the Reichsportsfeld. Apart from the central Olympiastadion, the most dramatic elements were the open-air swimming pool, the Dietrich Eckart open-air Grecian-style theatre, the enormous 11.2ha grass Maifeld, able to host 250,000 participants and spectators for mass Nazi May Day celebrations, and the 77m high Bell Tower at its western end. Around these were clustered smaller venues for the majority of Olympic events.

Also for the first time a landscape architect was involved. Professor Wiepking-Jurgensmann oversaw the mass transplanting of more than 40,000 mature trees up to 70 years old and 20m tall. The Official Report proudly boasts that not one of these trees failed to survive.

Contrary to a widely held belief, the new stadium was not built on the foundations of the old Deutsches Stadion. As the Official Report makes clear, the new stadium was shifted nearly 150m to the east to ensure the symmetry of the north/south axis of the Reichsportsfeld, and to free up space for the Maifeld.

A classic oval planform was chosen, with axes of 304m x 230m. Again, advantage was taken of the sandy soil conditions, with the inner bowl being excavated to a depth of 13.7m. The upper bowl rose 16.5m above ground level, creating a total of 71 tiers of seats in two sections, stepping downwards in a gentle parabolic curve from the top. A gallery 1m deep by 2.2m wide separated the spectators from the 400m red cinder running track and allowed officials to move freely without distracting the spectators.

In situ reinforced concrete was the main structural material. To meet Hitler's requirement, this "Modernist" structure was hidden away behind a cladding of Franconian shell limestone. Only native German materials were allowed on the Reichsportsfeld project: in all some ten different types of stone were used.

Berlin. Reichssportfeld. Stadion

The austere neo-Classic façade of the stadium was said to express the National Socialist vision of a new Germany.
SOURCE: IOC

Berlin's 1936 Olympic stadium was designed to promote National Socialism.
SOURCE: IOC

At the western end twin towers flanked the Marathon Gate, creating a gap in the seating bowl through which the Bell Tower was visible. There was a stone platform there, flanked by stone steps leading down into the central arena, on which the Olympic Cauldron was located. For the first time the Flame was fed by liquid propane gas.

The neoclassic architecture of the stadium with its clear geometry and austere façade was impressive in its time, and expressive of the Nazi vision of a new Germany. To the Nazis natural stone symbolized "indomitable German strength and the enduring nature of the National Socialist ideology."[8]

A network of tunnels under the stadium and much of the Reichsportsfeld, some dating back to the days of the Deutsches Stadion, allowed segregated access for VIPs and athletes. Ironically, Marathon competitors entered the stadium through a dedicated tunnel, rather than through the Marathon Gate itself.

Legacy

During the Second World War the stadium escaped serious damage, although other buildings on the Reichsportsfeld were less fortunate. In the immediate post-war years the British Army occupation forces annexed the complex as their headquarters, repairing and adapting buildings to their own requirements, and using the Maifeld as the venue for the annual celebration of the Queen's Official Birthday.

Derelict and fire-damaged, the Bell Tower was demolished in 1947. From 1963 the main stadium hosted Bundesliga football matches and became home to local football team Hertha BSC. The British Army left after nearly 50 years following the reunification of West and East Germany in 1990.

Post-reunification there was considerable debate over the future of the Olympiastadion, seen by many as a symbol of the Nazi era. It was finally decided to keep it, and a renovation programme began in 2000.

Some 90,000m³ of sand was excavated to lower the central field by 2.65m to meet the latest international specifications for football and athletics stadia. The lower tier of seating was rebuilt at a different angle of inclination. Obvious Nazi symbolism was removed, and the grandiose "sections of honour" accommodation for VIPs reduced in size. The limestone cladding panels were carefully dismantled and cleaned before reassembly.

Twenty steel columns were installed to support a translucent acrylic roof extended to 37,000m². Spectator capacity was reduced to 74,745, all seated, making it the largest stadium in Germany for international football matches. At the western end there was a gap left in the roof, which allowed the original view through the Marathon Gate to the Bell Tower (reconstructed in 1962) to be maintained. The roof has attracted criticism: it is said to be architecturally insensitive, expensive to maintain and providing little shelter from the elements in windy conditions.[9]

The original limestone cladding has been renovated and retained.
SOURCE: CREDIT SENA OZFILIZ

The Olympic stadium today.
SOURCE: CREDIT SENA OZFILIZ

In 2006 it hosted the football World Cup final. Hertha BSC are still based there, and it has been the venue for numerous international football matches, athletics championships and concerts.

Notes

1. Goldblatt, David (2016) *The Games: A Global History of the Olympics,* New York: W.W Norton, p. 17.
2. "The Nazi Olympics Berlin 1936", United States Holocaust Memorial Museum. www.ushmm.org
3. "The facade of hospitality", United States Holocaust Memorial Museum.
4. Clay Large, David (2007) *Nazi Games: The Olympics of 1936,* New York: W.W Norton, p. 58.
5. https://historynewsnetwork.org/article/571
6. www.die-fans.de/fussball/stadien/stadien-liste/dasstadion/,Deutsches+Stadion,147,,,,,northeast
7. Van der Vat, Dan (1997) *The Good Nazi: The Life and Lies of Albert Speer,* London: Weidenfeld & Nicolson, ISBN 978-0-297-81721-5.
8. Inglis, Simon (1990) *The Football Grounds of Europe,* London: Willow Books.
9. Ibid.

CHAPTER 15

LONDON
UK, 1948

Background

London was originally awarded the 1944 Summer Games at an IOC meeting in London in June 1939, not long before the outbreak of the Second World War in September of that year. There had been seven other original bidders: Athens, Budapest, Detroit, Helsinki, Lausanne, Montreal and Rome. These plans were inevitably abandoned, as were the 1940 Games, which had been transferred from Tokyo to Helsinki in 1937 (see Chapter 16).

When the IOC reassembled in 1946 it had to consider bids from Baltimore, Lausanne, Los Angeles, Minneapolis and Philadelphia as well as London. There were many who doubted war-torn London's ability to stage the Games and the country's ability to fund it. Food was still rationed, much of the city was in ruins. Even as a bid was prepared, many thought it would be more sensible to let the Games go to the United States, which had suffered far less physical and economic damage.

Eventually a postal vote of IOC members was held, and London came out on top. IOC president J. Sigfrid Edstrom was very influential in the decision. He said "Who was going to deny the gift of an Olympic Games to the city who had defied the might of Hitler's waves of bombers and rocket missiles?"

What became known as the "Austerity Games" or even the "Ration Book Olympics" was a triumph of improvisation and collaboration. No new venues were built; there was no Olympic Village. Competitors were issued with ration books – which did allot them the 5,647 calories normally only given to dockers and miners, as opposed to the otherwise universal 2,600 calories. Male athletes were accommodated in hastily refurbished army temporary buildings, women were housed in local colleges. For the first time, however, Olympic swimming events took place under cover, in the Empire Pool nearby, opened in 1934.

Germany and Japan were not invited: the USSR was invited, but chose not to send a team. A record 59 nations sent 3,714 men and 390 women to compete in 19 sports.

Fanny Blankers Koen, a 30-year-old mother of two from the Netherlands, shattered many misconceptions when she won three individual and one team athletic golds. As "the Flying Housewife" was also the world record holder in the high and long jumps she might have won more – had the rules not stipulated that no athlete could compete in more than three individual events.

These Summer Games turned out to be the most inexpensive and unpretentious of the 20th century. After the cynically propagandist Berlin Games of 1936, it was felt that they were a triumph of the spirit of fair play.

Stadium origins

London's 1908 Olympic stadium at White City had survived the war unscathed. Although mainly devoted to "non-Olympic" events such as greyhound racing and motorcycle speedway, the stadium did have a 440yd (402.3m) cinder running track, which had replaced the original, "three laps to the mile" track in 1931 (see Chapter 8). It had also hosted the second British Empire Games and the fourth Women's World Games in 1934, and was the venue for the annual Amateur Athletics Association Championships.

Despite all this, the London organizing committee preferred what was then known as the Empire Stadium, further from the centre of London and located on the old pleasure grounds of Wembley Park. The

An understated poster for the "Ration Book Olympics".
SOURCE: IOC

park was the brainchild of Victorian railway entrepreneur Sir Edward William Watkin, the chairman of the newly built Metropolitan Railway (now the Metropolitan Line, part of London Underground), and was intended to draw visitors from Central London. These would travel on his railway, Watkin believed, and alight at a specially built new station.

Among the many attractions at Wembley Park were football and cricket fields and a golf course. The main attraction, however, was to be the Great Tower of London, a vainglorious structure taller even than the Eiffel Tower. Construction actually began in 1892, but had only reached a height of 47m before funds ran out. What was derisively dubbed "Watkin's Folly" or the "London Stump" was eventually demolished in 1904.[1]

When plans for the 1924–25 British Empire Exhibition were drawn up in 1920, the Great White City, the venue for the 1908 Franco-British Exhibition (see Chapter 8) was rejected as a site for reasons that remain obscure. Instead, Wembley Park was chosen, despite the vehement objections of the local authority.

Along with an ornamental lake and pleasure gardens, and hundreds of commercial pavilions, there were to be 16 major pavilions for member countries of the empire, plus the three massive Palaces of Engineering, Industry, and the Arts, along with the almost equally massive HM Government building.[2]

Work started first, however, on a new 125,000 capacity stadium located on the site of Watkin's Folly. Like all the major Exhibition structures it was to be constructed largely of "ferroconcrete" – in situ reinforced concrete. Some in the UK saw this as a daring innovation, despite its growing acceptance in other countries.

Design team

Architects: John William Simpson (1858–1933) designed many public buildings, including the famous Roedean School for girls. In 1910 he took the young Maxwell Ayrton (1874–1960) into partnership.

Structural engineer: Sir Owen Williams (1890–1969) was best known for his pioneering work with reinforced concrete. Among his most high-profile projects were the Empire Pool, and what was quickly dubbed "Spaghetti Junction" – Gravelly Hill Interchange on the M6 motorway in the UK's Midlands.

Design

Around 250,000t of earth had to be removed to create the stadium "bowl". Some 25,000t of concrete, 2,000t of steel, 22.5km of precast concrete beams for the terraces and 500,000 rivets went into the stadium structure. Construction took just 300 working days, at a cost of £750,000 – just £30m at today's prices.

A series of rounded arches formed the exterior wall, with the iconic Twin Towers at the entrance side of the stadium. Overall, the external shape followed the contours of the site, which helped to give it a sense of height.

In 1923, when it hosted the FA Cup Final for the first time, it was unquestionably the largest and grandest soccer venue ever built.[3] The final itself was a near disaster. An estimated 250,000+ spectators squeezed into the stadium and overflowed onto the pitch. Mounted police managed to clear the playing surface and the match eventually went ahead. This was probably the largest audience ever to attend a sporting fixture. All subsequent events were ticket-only.[4]

Unfortunately the Exhibition was a financial failure, and in 1925 the demolition men moved in. Most of the pavilions were torn down: the stadium itself was also slated for demolition, and was only saved by the opportunism of entrepreneur Arthur Elvin. He believed those who had dismissed the stadium as financially unviable were short-sighted, and purchased it for £127,000 – around £7m at today's prices.

Elvin's first move was to install a greyhound racing track, which proved to be a financial goldmine for many decades. Motorcycle speedway followed in 1929, and the stadium continued to host FA Cup Finals and international soccer matches – but only between England and Scotland.

Wembley Stadium during the 1948 Summer Games.
SOURCE: IOC

Upgrade for the Summer Games

In the spirit of the "Ration Book Olympics" only essential changes were made to the stadium. There had been no athletics staged there for more than 20 years, so the major challenge was to rip up the greyhound racing track and replace it with a cinder 400m running track. The electric hare equipment, fencing and arc lights had also to be removed, all within a three-week window.

A giant scoreboard was erected at the eastern end of the stadium, operated manually during the Games by a party of naval ratings. A cauldron for the Olympic Flame was installed, and a somewhat basic and rickety Royal Box.

Legacy

Over the next eight decades what became known as Wembley Stadium also became an international icon, symbolized by the 126m high "Twin Towers". Best known as a soccer

After the second-ever Olympic Torch relay, the Olympic Flame arrives at Wembley Stadium, carried by 400m runner John Mark.

SOURCE: CREATIVE COMMONS – CREDIT NATIONAL MEDIA MUSEUM

venue, the stadium hosted many soccer cup finals and international matches, including the 1966 World Cup final, which England won for the first and only time.

Rugby league, American and Gaelic football also took place, while greyhound racing and motorcycle speedway continued to attract large crowds. It was also the host stadium for the global Live Aid concert in 1985.

Like virtually all multi-purpose stadia, Wembley was far from ideal for field sports. Spectators could be a very long way from the action. As the stadium aged its limitations became ever more obvious. Even a major renovation in 1963, which mainly involved converting it to an all-seated venue with a capacity of 82,000, had only limited success.

Sightlines and seat dimensions proved to be inadequate. Little could be done to improve the notoriously spartan toilet and refreshment facilities "below stairs". In 2000 the decision was taken to replace it with a completely new and more intimate stadium designed for field sports only, primarily association football, but also rugby league, rugby union and American football. It is also suitable for music concerts and conferences.

There was vociferous opposition to the demolition of the much-loved original stadium, and a high-profile campaign to save the Twin Towers at least. In the end, however, it was found that the towers were not the lightweight structures assumed in the rescue proposals, and demolition was the only realistic option. The top of one of the towers was saved and erected as a memorial in a nearby park.

The original Wembley Stadium's iconic Twin Towers were demolished in 2003.
CREDIT: SIMON INGLIS

The new Wembley Stadium with its iconic arch, designed by Populous and Foster + Partners.
SOURCE: CREATIVE COMMONS – CREDIT WIKIOLO

Notes

1. https://web.archive.org/web/20090502015057/http://www.wembleystadium.com:80/StadiumHistory/historyIntroduction
2. https://web.archive.org/web/20160427065725/http://www.engineering-timelines.com/scripts/engineeringItem.asp?id=395
3. Inglis, Simon (1990) *The Football Grounds of Europe*, London: Willow Books.
4. Matthews, Tony (2006) *Football Firsts*, Capella. ISBN 1-84193-451-8.

CHAPTER 16

HELSINKI
FINLAND, 1952

Background

Imperial Japan's invasion of China in 1937 led to the IOC stripping Tokyo of the honour of hosting the 1940 Summer Games. Helsinki had campaigned for the 1940 Games but had been ignored at the IOC meeting in Berlin in 1936. Campaigns by Barcelona and Rome had been similarly unsuccessful: however, it was Helsinki that was awarded the Games by default in 1937. Inevitably the Games had to be cancelled when the Second World War broke out.

In 1947 Helsinki was the clear winner when voting for the host city for the 1952 Summer Games took place in Stockholm. Minneapolis and Los Angeles tied for second place ahead of Amsterdam; Detroit, Chicago and Philadelphia attracted few votes.

Helsinki is the furthest north the Summer Games have ever been held. Until the Beijing Olympics of 2008 Helsinki was the Games at which the most world records were broken.

Among the 13 nations making their debuts at the 1952 Games were the Soviet Union, the People's Republic of China and Israel. Germany and Japan were allowed to send teams again after being barred from the 1948 Summer Games, but the German Democratic Republic (East Germany) refused to participate in a single united German team with the Federal Republic of Germany (West Germany).

Saarland, a short-lived German protectorate under French control since 1947, sent a team to the Olympics for the first and only time. In all, 69 countries were represented by 4,932 athletes, 521 of them women, who competed in 17 sports.

Origins

In the run-up to Helsinki's bid for the 1940 Olympics a Stadium Foundation was established as far back as 1927, and a design competition held in 1933. The chosen site was in the Eläintarha park, some 2km from the centre of Helsinki and close to an existing football stadium.

There was surprise and dismay when Tokyo won the bidding in 1936, beating not only Helsinki but also Barcelona and Rome. In Helsinki, stadium construction had begun in February 1934, and work on the stands was well underway by the time the IOC announced Tokyo as the chosen venue.

Nevertheless, construction went ahead and the stadium was inaugurated on 12 June 1938.

Helsinki's Olympic Stadium in 1938, ready for the cancelled 1940 Summer Games.
SOURCE: PUBLIC DOMAIN

The eastern stand was unfinished at the time, and was dubbed the "Olympic hole". Local cartoonists had a field day.

The stadium was also scheduled to host the 1943 Workers' Summer Olympiad,[1] also cancelled.

Design team
Architects Yrjö Lindegren (1900–1952) and Toivo Jäntti (1900–1975) won the 1933 design competition. Lindegren would go on to win an Olympic gold medal for architecture at the 1948 Olympics in London, the last time Olympic art competitions were held.

Structural design is credited to local practice Jernvalli and Laakso.[2]

Design
Chicago architect Louis Sullivan famously coined the phrase "form ever follows function" in 1896, sparking a controversial architectural movement known as Functionalism. By the 1930s it had been embraced particularly in Central and Northern Europe, where it was interpreted in the form of buildings that eschewed ornamentation and architectural whimsy.

Lindegren and Jäntti's original design for the Helsinki Olympic Stadium is often hailed as the most successful Functionalist building ever constructed. In later years it was also frequently

Opening Ceremony of the 1952 Summer Games.
SOURCE: IOC

described as the most beautiful of all Olympic stadia. Its clean, uncluttered lines took full advantage of the design flexibility offered by reinforced concrete construction, then by far the most popular structural option in Nordic countries.

An iconic tower is the most outstanding feature. Its height, 72.71m, commemorates the winning javelin distance thrown by 1932 Finnish Olympic gold medal winner Matti Järvinen.

When it first opened, the stadium had a capacity of only 26,000 spectators. "The western grandstand was fully covered by a concrete roof of tapered profile that is cantilevered over a single row of steel columns with conical concrete capitals."[3]

Once the 1940 Games were finally awarded, the eastern stand – the "Olympic hole" – was completed and additional wooden stands added to bring spectator capacity up to 62,000.

Winning the vote to host the 1952 Games triggered a major renovation and extension, supervised by Lindegren. New temporary stands were added, increasing capacity to 70,000. During the Games the stadium hosted athletics and equestrian events, and the football finals, as well as the Opening and Closing Ceremonies.

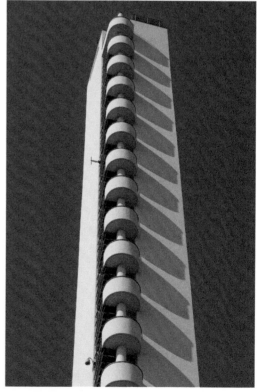

Helsinki's Olympic Stadium Tower is 72.71m high – the javelin gold medal winning throw by a Finnish competitor at the 1932 Summer Games.
SOURCE: CREATIVE COMMONS – CREDIT ODIN

Athletes compete at the 1952 Summer Games.
SOURCE: IOC

Two Olympic flames were lit for the Games: one burned on top of the tower, one, more visible to spectators, in a cauldron in the infield. The flame was brought into the stadium at the opening ceremony by Paavo Nurmi, the winner of nine gold and two silver medals for athletics at previous Olympics. A statue of Nurmi, the "Flying Finn", was erected in front of the stadium that same year.

Legacy

Removing the temporary stands left the stadium with a capacity of 50,000 spectators. The central area was converted into a giant ice rink in 1957 to allow the Bandy World Championship to take place.[4]

In 1971, with the running track enlarged from six to eight lanes and resurfaced, the stadium hosted the European Athletics Championships for the first time. It hosted them again in 1994 and 2012.

A temporary 5,000-seater stand was installed in 1983 for the first World Athletics Championships. Between 1991 and 1994 the stadium underwent a major renovation and refurbishment. A roof appeared over the east stand in time for the 2005 World Athletics Championships. Spectator capacity was now 40,000, making it still the largest stadium in Finland.

Apart from being the home stadium of the Finland national football team, it has hosted many concerts over the years. The original Functionalist building still survives at the core of the modern stadium, which is seen as a major asset to Helsinki and, indeed, the whole of Finland.

In 2016 the stadium closed and a three-year major upgrade and modernization project began. This will see roofs over all spectator areas, a renewal of the entire arena, and more indoor sports facilities.[5]

Notes

1. Launched as an alternative to what were seen as Games dominated by an aristocratic and privileged elite, the International Workers' Olympiads were a series of Summer and Winter Games running from 1925 to 1937. Supported by the trade unions and social democratic parties, these events encouraged mass participation. Some 100,000 competitors turned out at the 1937 Summer Olympiad in Vienna, watched by 250,000 spectators – considerably better totals than for the 1932 Summer Games in Los Angeles.
2. https://www.archinform.net/projekte/3148.htm
3. Gordon, Barclay F. (1983) *Olympic Architecture: Building for the Summer Games*, New York: John Wiley. ISBN -13: 978-0471060697.
4. Only three nations competed at these championships: the Soviet Union, Sweden and Finland. Bandy is a form of ice hockey that originated in London in 1875. It is played outdoors on a frozen pitch approximately the same size as a soccer pitch.
5. www.stadion.fi/en/prostadion

CHAPTER 17

MELBOURNE
AUSTRALIA, 1956

Background

No fewer than ten cities bid for the 1956 Summer Games in 1949. Six were from the USA: Detroit, Chicago, Los Angeles, Minneapolis, Philadelphia and San Francisco. Montreal and Mexico City were also in the race. The two Southern Hemisphere cities, Buenos Aires and Melbourne were initially seen as rank outsiders, as the reversal of seasons would see the Games held during the Northern Hemisphere winter. This would adversely affect athletes normally accustomed to competing during the Northern summer, it was alleged.

Nevertheless, right from the first round of voting at the IOC session in Rome, Melbourne took the lead, with Buenos Aires and Mexico City still in contention. The fourth round was a two-horse race with a close finish, Melbourne triumphing over Buenos Aires by just one vote.

Problems surfaced almost as soon as the victory was announced. It was revealed that strict Australian quarantine laws would prevent Melbourne hosting the equestrian events: eventually they were held in Stockholm, Sweden, instead (see Chapter 9). Australian politicians bickered and squabbled over the financing of the Games. There were equally frenetic arguments and delays over the siting of the main stadium, so much so that the IOC seriously considered transferring the 1956 event to Rome, which was well ahead with preparations for the 1960 Summer Games.

Much to Melbourne's relief the problems were sorted out and the Games opened on time. Even the challenge of the Olympic Flame relay was overcome, with the Flame arriving in the North of Australia by plane and then travelling down the east coast to Melbourne. En route it passed through Sydney, but its actual arrival in Sydney was almost overshadowed by students bearing a home-made torch.

This was another Summer Games to be hit by boycotts. Egypt, Iraq and Lebanon withdrew over the Israeli, British and French invasion of Egypt in 1956. In the same year, after the Hungarian uprising was crushed by the Soviet Union, Spain, Switzerland, the Netherlands and Cambodia announced they would not be attending. Finally, less than two weeks before the Opening Ceremony, the People's Republic of China pulled out in protest against the inclusion of the Republic of China (Taiwan).

In total, 67 countries attended. Competitor numbers were down significantly from the 1952 Helsinki Games, with just 2,938 men and 376 women participating. Including the six equestrian events held in Stockholm – where an additional 158 competitors took part – there were 17 sports in total.

A new Summer Games tradition was started when John Wing, an Australian-born citizen of Chinese descent, wrote anonymously to the IOC to suggest that at the Closing Ceremony competitors should be allowed to mingle together rather than marching in as national teams. This was immediately adopted, earning the Melbourne Games the soubriquet of the "Friendly Games".

Stadium origins

Of all the disputes that followed the award of the Games, perhaps the most serious was over the location of the main stadium. Powerful groups lobbied for a number of different options, and the arguments dragged on until the very future of the Melbourne event was in doubt. In the end Australian prime minister Robert Menzies stepped in and forced through a compromise on both the stadium and the funding.[1]

With time running out, the option of building a completely new stadium no longer existed. Instead, the historic Melbourne Cricket Ground (MCG) would be refurbished and modernized.

Cricket was first played on the site in 1853, and in 1877 it hosted what was later acknowledged as the first Test match between England and Australia. The land had been an

This 1877 grandstand was one of the first structures at the historic Melbourne Cricket Ground.

important Aboriginal camping ground until 1835, when it was appropriated as a grazing area for military horses. Although the site was not quite level, the slope was not so great as to make it unsuitable for cricket, while steep enough to ensure good drainage.

It is located close to the city centre, with excellent car parking facilities and good access by rail and road.

Design

There was never a masterplan for the MCG; it simply evolved over more than ten decades.[2] Starting with a wooden members' stand in 1854, stands were added, replaced with brick or concrete structures, and replaced in their turn. Spectator capacity grew steadily. By 1937, following the completion of the Southern Stand, the ground could officially accommodate 84,000 seated and 94,000 standing, and at least two tiers of seating ringed most of the arena.[3]

In practice, despite its name, it was Australian rules football that attracted the largest crowds to the MCG. Football was first played there in 1879, and continues to be the most popular sport to this day. The ground has also hosted intermittent appearances of both forms of rugby, but both failed to attract substantial interest. Soccer, however, fared somewhat better.

By 1914 the evolution of the MCG was well underway.
SOURCE: PUBLIC DOMAIN

The Scene on Opening Day.

The U.S.S.R. Team can be seen on the right.
The U.S.A. Team is in front of the Choir in
the centre of the photograph.   The Australian
Team is entering the Stadium on the bottom lef
The Swimming, Soccer, Hockey and Cycling Arena
can be seen beyond the Main Stadium.

The long evolution of the Melbourne Cricket Ground is evident in this aerial view of the
Opening Ceremony of the 1956 Summer Games.
SOURCE: IOC

Olympic upgrade

For the Olympics it was decided to replace the 9,000-seat Grandstand, which dated from 1884,
with a new three-tiered 40,000-seat concrete framed stand, variously known as the Northern or
Olympic Stand, which would increase capacity to a nominal 120,000.

A more radical reshaping was needed for the arena itself. The fall of 2.4m from north to
south exceeded the official limits for athletics – maxima of 1 in 1,000 in length and 1 in 100
in width. It had to be totally regraded. A seven-lane 400m cinder track was laid down. The
temporary nature of the track, which was removed immediately after the Games, may explain
why there were many complaints from those who competed on it. The inside lane cut up very
badly, to the point that some distance runners were forced to use outer lanes instead.

Legacy

Described in 2003 as "a shrine, a citadel, a landmark, a totem" that "symbolises Melbourne
to the world,"[4] the MCG has continued to evolve almost constantly since the 1956 Games. In
1968, following the completion of a new stand, the MCG reached its maximum capacity.

As this 2017 image illustrates, Australian Rules football is one of the few sports compatible with an oval athletics stadium.
SOURCE: CREATIVE COMMONS – CREDIT FLICKERD

An official world record for a sporting event was set when 121,696 spectators attended the 1970 Australian rules football final, although this figure was topped by the estimated 143,000 who attended a Billy Graham evangelical meeting. (The "White Horse" Cup final at Wembley in 1923 had an estimated attendance of 250,000+, but no official records were kept.)

What are claimed to be the world's highest lighting towers at 85m made their appearance in 1985, and there were major refurbishments and reconstructions before the 1992 cricket World Cup and the 2006 Commonwealth Games. Spectator capacity is now just over 100,000, making the MCG the largest stadium in the Southern Hemisphere and the tenth largest in the world.

Unlike many other major stadia, the MCG has no quiet period. Cricket dominates the summer, Australian rules football the winter, yet despite this heavy usage the arena is still natural grass. And time and space has also to be found for minority sports, major concerts and Papal visits.

Notes

1. Lewis, Wendy, Simon Balderstone and John Bowan (2006) *Events That Shaped Australia*, Frenchs Forest, NSW: New Holland, pp. 212–217. ISBN 978-1-74110-492-9.
2. "Melbourne Cricket Ground, Victorian Heritage Register (VHR) Number H1928, Heritage Overlay HO890". Victorian Heritage Database. Heritage Victoria.
3. "MCG Facts and Figures" (2009). Melbourne Cricket Ground.
4. Baum, Greg (2003) "MCG voted as one of the seven wonders of the sporting world", *The Sydney Morning Herald*, 24 September.

CHAPTER 18
ROME
ITALY, 1960

Background

A major eruption of Mount Vesuvius in 1906 devastated much of the city of Naples. Government funding for the planned 1908 Summer Games in Rome had to be diverted to the reconstruction of Naples, and so the Games were hastily transferred to London (see Chapter 8).

Rome bid for the 1940 Summer Games, but lost out to Tokyo. In Italy it was assumed that Rome would get the 1944 Games and, under the Fascist regime of Mussolini, there was a major construction programme aimed at creating a world-leading sports complex in the city, the Foro Mussolini. The Second World War intervened, and the project was left partially incomplete.

In 1955, however, at the IOC session in Paris, France, Rome scored a third-round victory over Lausanne, Detroit, Budapest, Brussels, Mexico City and Tokyo. This was seen as a historic moment, as the order to end the Ancient Games in Greece had come from Imperial Rome in AD 394.

There was much more public and official enthusiasm for the Summer Games than there had been in 1905. The government saw the Games as an opportunity to modernize and revitalize the historic city, and to shrug off the negativity that followed Italy's defeat in the Second World War. Funding came largely from the hugely popular Totalcalcio football pool system.

New hotels were built, there were new road links and a new municipal water supply. For obvious reasons the Foro Mussolini had already been rebranded as the Foro Italico. New facilities were added, including two sports palaces designed by Pier Luigi Nervi, the leading engineer/architect of his day, and both spectacular examples of the potential of reinforced concrete.

Rome's long history was celebrated by a number of events held in venues dating from Imperial Rome. Gymnastics took place in the ruins of the Baths of Caracalla; Olympic wrestling was staged in the ruins of the 4th-century Basilica of Maxentius.

There were no official boycotts, although the sole competitor from Suriname withdrew at the last minute. In all 83 nations sent 611 female competitors and 4,727 males to take part in 17 sports. South Africa made its last appearance as an apartheid state, only reappearing in the 1992 Barcelona Games.

Two years before the first television relay satellites were launched, videotapes of the day's events were flown across the Atlantic to New York, from where they were broadcast on national TV network CBS. Because of the time difference between Rome and the United States it was often possible to broadcast daytime events on the "same" day they took place.

For the first time ever, the main Games were followed a week later by what became known as the Paralympic Games, although they were originally billed as the 9th International Stoke Mandeville Games (see Chapter 4). There were 21 national teams comprising some 350 athletes with spinal injuries, competing in eight sports. No events took place in the Stadium Olimpico.

This may have seemed a modest beginning, but these first Paralympics were to be acknowledged as having world significance. Sir Ludwig Guttmann, generally known as the "Father of the Paralympics" (see Chapter 4), stated in his Closing Ceremony speech: "The vast majority of competitors and their escorts have fully understood the meaning of the Rome Games as a new pattern of the re-integration of the paralysed into society, as well as the world of sport."[1]

Located in a splendid natural amphitheatre, Rome's Stadio Olimpico offered little protection to spectators during the 1960 Summer Games.
SOURCE: IOC

Stadium origins

Located to the north of the city centre, what was originally dubbed the Stadio dei Cipressi was the centrepiece of the Fascist-era Foro Mussolini. Construction of the first phase began in 1927. The original design was a straightforward oval stadium, in a spectacular natural amphitheatre formed by tree-covered hills.

Seating was to be on grass-covered terraces, but this was changed to masonry construction at an early stage. In 1937 construction of an upper tier of seating began, but was abandoned in 1940.

Work restarted in 1953. There were plans for an ambitious structural extension that was never realized due to shortage of funds and concerns about environmental impact. Nevertheless, capacity now reached 100,000 spectators, giving rise to a new name, Stadio dei Centomila. Tiles of classic travertine marble clad the concrete structure internally and externally.

Design team

Turinese engineer Angelo Frisa and architects Enrico Del Debbio and Mario Moretti were responsible for the original design.

In 1950 responsibility for completing the original stadium fell to engineer Professor Carlo Roccatelli, a member of the Superior Council of Public Works. He died in 1951, and was replaced by the architect Annibale Vitellozzi.[2]

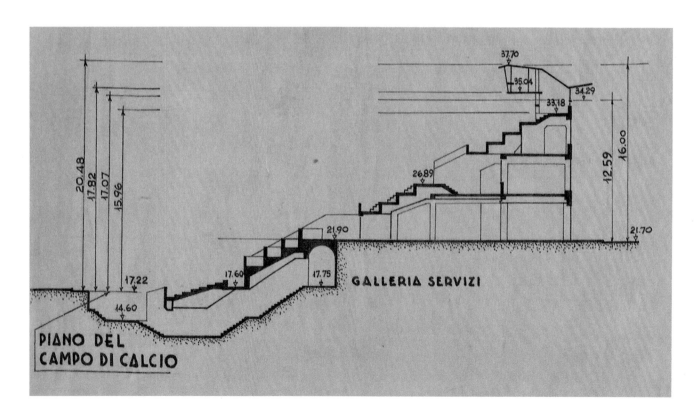

Design

A classic symmetrical oval with external dimensions of 319m on the major axis and 186m on the minor axis, the field level of the Stadio dei Cipressi was 4.5m below external road level. The transformation into the Stadio dei Centomila saw tiers of seating up to 20.5m above field level.

For the Summer Games spectator capacity was reduced to 65,000, mainly by removing the field level seating. There was another name change, to the Stadio Olimpico, which endures to this day.

The spectator terraces were largely open and uncovered. Low partitions of tempered glass divided the seating areas into sections, allowing a visual flow around the interior of the stadium. A 2m wide and 1.9m deep trench was cut around the entire perimeter of the infield. This had the advantage of preventing spectator invasions.

During the Games seating for journalists was increased to more than 1,100, and there were 40 aluminium and glass cabins for television commentators. Four lighting towers delivered 250 lux of illumination at field level.

As was the norm at the time, the running track was red cinder. Infield drainage and irrigation was upgraded, although a week after the Closing Ceremony a rainstorm turned the infield into a lake.

This image of the Opening Ceremony clearly shows the deep trench that encircles the entire infield.
SOURCE: CREATIVE COMMONS, CREDIT – ALEX DAWSON

The stadium today in its urban setting, displaying its fibreglass membrane roof that was added during a major renovation in 1987.
SOURCE: CREATIVE COMMONS, CREDIT – DOC SEARLS

Legacy

Subsequent to the Summer Games the Stadio Olimpico hosted a number of athletic championships, and the 1977 and 1984 European Cup finals. Then, when Italy won the right to hold the 1990 FIFA World Cup, the stadium was chosen as the venue for the final, and a major renovation began in 1987.

Ultimately almost all the original stadium was demolished, apart from one stand. A new concrete main structure arose, while a free-standing triangulated steel tube superstructure supported a tensile polytetrafluoroethylene (PTFE) coated fibreglass membrane roof. Spectator capacity was now 82,911.[3]

In 2008 another major renovation was deemed necessary. Seating capacity was reduced to 73,261; spectator restrooms and toilets were increased. The stadium is now home to the Serie A football clubs of Lazio and Roma.

Notes

1.   https://www.paralympic.org/rome-1960
2.   www.worldstadiums.com/stadium_menu/architecture/stadium_design/roma_olimpico.shtml
3.   www.birdair.com/projects/rome-olympic-stadium

CHAPTER 19

TOKYO

JAPAN, 1964

Background

When Imperial Japan invaded China in 1937 the IOC stripped it of the honour of staging the 1940 Summer Games, for which Tokyo had been the sole bidder the previous year. Instead, Helsinki, Finland, was nominated, but these plans were also abandoned due to the Second World War.

Just 15 years after the war the IOC met in Munich to select the host city for the 1964 Summer Games. Tokyo was up against Brussels, Detroit and Vienna, but cruised to victory in the first round of voting. "The Olympic Games have arrived in Asia at last" was the general verdict.

By this time Japan was enjoying a sustained economic boom – but Tokyo's infrastructure was struggling to cope with the massive influx of new workers from the countryside. Much of the city had been devastated by US bombing raids and hastily rebuilt in a piecemeal fashion. The Games were seen as a once-in-a-lifetime opportunity to transform and modernize what was becoming one of the largest cities in the world, with more than 10m inhabitants.

A major urban upgrade was planned, centred on the Olympic sites. There were to be new highways and subway lines, a new international airport and a much-needed modern sewerage system. New sports centres and new parks were accompanied by commercial and residential developments; new energy, growth and prosperity arrived in the city.

At the heart of the Games were a number of new Olympic projects, the architecture of which received international accolades. British Olympic gold medallist turned sports journalist Christopher Brasher eulogized architect Kenzo Tange's Yoyogi National Gymnasium, host to the swimming events, as a masterpiece, "with its two great curtains of curving concrete, draping rather than roofing the water."[1] Other Kenzo Tange designs stimulated architecture all over the world.

The massive construction programme was completed on time, albeit at the cost of more than 100 deaths on Olympics-related projects and wide-scale disruption to the city centre. Some 200,000 stray cats and dogs were rounded up and euthanized before the Games began.[2]

For the first time ever, Olympic events were broadcast live to the United States via satellite television. There were also the first ever colour TV broadcasts, although these were only available in the Greater Tokyo area. Computers were utilized on a much larger scale than ever before, and Seiko took over timing responsibility from Omega, a complete break with tradition.

Political clashes over the issue of the "Games of the New Emerging Forces" (GANEFO)[3] saw North Korea, China and Indonesia boycott the Tokyo Games. The IOC also finally banned apartheid-era South Africa. Nevertheless a record 93 countries sent 678 women and 4,473 men to compete in 19 sports. There was universal praise for the impeccable organization of the Games.

Stadium origins

There had been a multi-use 66,000-spectator capacity facility on the central Tokyo Meiji National Park – otherwise known as the Outer Garden of the Meiji Shrine – since 1924, when the Meiji Jingu Gaien Stadium opened. It hosted the 1939 Far Eastern Games and was due to be the main stadium for the projected 1940 Summer Games, but was demolished in 1956 to be replaced by a new National Stadium with a capacity of 52,000, mostly seated.

This opened in 1958, and played host to the Third Asian Games that same year. More than 1,800 athletes from 20 countries took part. There was excellent access to the stadium, particularly by public transport, so the decision was taken to upgrade and extend it for the Olympics.

Tokyo's National Stadium, seen here during the Opening Ceremony for the 1964 Summer Games, is one of the most beautiful ever built.
SOURCE: IOC

Design team
Upgrade architect: Matsuo Katayama, Kanto District Bureau of Construction Ministry. No structural engineer is credited in the Official Report.

Design
A near 10m cross fall made the site a challenging one. Original construction was primarily in reinforced concrete with some structural steel elements. Initially it was hoped to increase spectator capacity to 100,000, but the practical limit turned out to be around 75,000.

To increase capacity a new crescent-shaped grandstand was added, with the highest "bleachers" 31m above the infield. A new 200m long underground passageway was built, to connect the infield with the locker and changing rooms, and the athletes' waiting rooms.

Overall, the form was quite beautiful. Circular in plan, with asymmetric shaping of the seating around the bowl, the stadium had a covered grandstand on one side.

Supplementary artificial floodlighting was installed, mounted on four 50m high lighting towers. These helped maintain the light levels needed for colour television transmissions: 500 lux on average, 1,500 lux at goals, pits and circles.

For the last time at a Summer Games the eight-lane 400m track was made of traditional "cinders". Given Tokyo's frequently inclement weather this had to be particularly free draining, so, after a period of intensive testing, a material known as neo-H-brick was selected, and laid 30mm thick.

A related test programme was used to choose the grass variety for the turf installed in the 100 x 68.5m infield. To keep it in good condition there were 155 irrigation sprinklers scattered around the infield.

Other upgrades included a new electronic bulletin board capable of displaying up to 500 words. Another Olympics first was the use of photo finish equipment for the sprints.

The Olympic Flame burned in a conical cast-iron cauldron 2.1m in diameter and 2.1m deep, located at the top of the large asymmetric seating area.

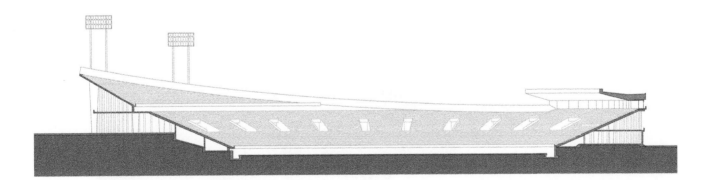

The 400m athlete Yoshinori Sakai lit the Olympic Cauldron at the 1964 Tokyo Summer Games.

Sports such as soccer and American football fit uneasily into an elliptical athletics stadium, as this image clearly shows.
SOURCE: CREATIVE COMMONS, CREDIT – NONE

Legacy

After the Olympic Games, the stadium hosted a number of major events. It was the home of the National Japanese Football team, and major football cup finals were held there. In 1991, the stadium hosted the World Athletics Championships. Concerts were a regular event, and it survived the 2011 Tohoku earthquake and tsunami unscathed.

When Tokyo was awarded the 2020 Olympic Games in 2013 (see Chapter 33), it was decided to demolish the stadium, and build a new one on the same site. A special funeral concert was held in the stadium before demolition commenced, and demolition was completed in May 2015.

Notes

1. Brasher, Christopher (1964) *Tokyo 1964: A Diary of the XVIIIth Olympiad*, Stanley Paul.
2. Whiting, Robert (2014) "Negative impact of 1964 Olympics profound", *Japan Times*, 24 October, p. 14.
3. GANEFO was established by Indonesia after the IOC suspended its membership for banning Taiwan and Israel from the 4th Asian Games held in Jakarta in 1962. In 1963, 51 mostly nominally socialist countries sent teams to the first and only GANEFO, held in Jakarta. The IOC decreed that athletes attending GANEFO would not be allowed to compete in the Olympic Games, but it did reinstate Indonesia before the Tokyo Games.

CHAPTER 20

MEXICO CITY
MEXICO, 1968

Background

It was a first-round victory for Mexico City when the IOC met in Baden-Baden, West Germany, in 1963. Mexico's capital easily beat off challenges from Detroit, Lyon and Buenos Aires, making it the first Latin American country to host the Summer Games.

There were many who doubted the wisdom of the IOC's decision. The Games would be held at an elevation of 2,240m (7,350ft) above sea level, by far the highest ever, where air pressure was significantly lower. And those who saw Mexico as a land of "slumber and mañana"[1] found it hard to believe that the country could organize an event as complex and challenging as the Summer Games, let alone afford it.

Trial athletic events held before the Games showed there was no significant threat to competitors' lives or health. Endurance athletes would see performances reduced: explosive events such as sub-400m sprints, jumps, leaps and throws would see sometimes dramatic improvements, thanks to "thinner air" – lower air resistance.

A budget less than 10 per cent of Tokyo's four years earlier meant the opportunities for creating new infrastructure were limited. There were few new venues. Even the Olympic Village was simply rented from a developer.

There was, however, the staging of an Olympics Cultural Festival or Olympiad, which ran for nine months before the Games began. More than 1,500 events were staged, including film and folklore festivals and musical and dance performances. Hundreds of local artists and performers took part, and major international artists were invited to contribute. These included poets Robert Graves and Yevgeny Yevtushenko, playwright Arthur Miller and jazz musicians Duke Ellington and Dave Brubeck.

Unfortunately these events were overshadowed by the political unrest in the run-up to the Games. Mexico had effectively been a one-party state for five decades, and the government's authoritarian policies were increasingly unpopular. It had always responded with extreme violence to any protests or anti-government demonstrations, and the Olympic year was to be no different.[2]

A series of marches and rallies throughout the summer attracted the predictable brutality from the security forces. In early October, with the Opening Ceremony just 10 days away, a rally in the Tlatelolco housing development turned into a massacre.

Some 5,000 soldiers and paramilitaries backed up by 200 small armoured vehicles fired on 10,000 demonstrators and bystanders for several hours. The actual number of dead has never been established: estimates range between 20 and 30 (local government-controlled media) to as high as 400.[3] As many as 1,000 were wounded, hundreds more were arrested, imprisoned and tortured.

Nevertheless, the Games went ahead as scheduled. In all, 781 women and 4,735 men from 112 countries competed in 18 sports. North Korea was the only country to boycott the 1968 Summer Games: for the first time, East and West Germany competed as separate teams. It was also the first Games to be transmitted worldwide in colour, and where tests for illegal doping were introduced.

There was controversy during the Games, when the black American gold and bronze medallists in the men's 200m gave Black Power salutes during the medal ceremony. Tommie Smith and John Carlos were subsequently banned from all future Olympics. On a more positive note, this was the first Summer Games at which the Olympic Cauldron was kindled by a woman, track and field athlete Norma Enriqueta Basilio de Sotelo.

Stadium origins

Opened in 1952 and then known as the Estadio Universitario, what became the 1968 Olympic Stadium was located on the main campus of the National Autonomous University of Mexico, in the southern part of Mexico City. The stadium bowl was scoured out of an ancient lava bed up to eight metres thick, with the excavated lava used to face an asymmetrical "saddle-shaped" berm.

Famous Mexican artist Diego Rivera undertook to decorate the external sloping lava façades with murals evoking the cult of the sports of the indigenous civilizations of Ancient Mexico.

Artist Diego Rivera sadly died before he could complete his dramatic mosaic mural around the entire circumference of the stadium.
SOURCE: CREATIVE COMMONS – CREDIT ELSA.ROLLE

Sadly, he died before he could complete the work, leaving behind just a dramatic mosaic mural above the main entrance on the east side. Overall, the stadium attracted many plaudits, and is still considered to be one of the most beautiful major stadia in the world.

Tunnels and ramps were dug out of the encircling berm to give access to the stands and seating areas. Concrete structures supported the VIP boxes. Original capacity was 65,000 spectators, who came initially to watch collegiate American football matches. The 1955 Pan American Games was the first major event to be staged here.

Design team

Original architects: Augusto Pérez Palacios, Jorge Bravo and Raúl Salinas Moro.

Design of the upgrades: Pedro Ramírez Vázquez (1919–2013) was an award-winning and prolific Mexican architect, a member of the IOC and the president of the Mexico City organizing committee. He designed the larger Azteca Stadium in Mexico City (opened 1966), which showcased his preference for concrete. Vázquez also served as minister of public infrastructure and human settlements.

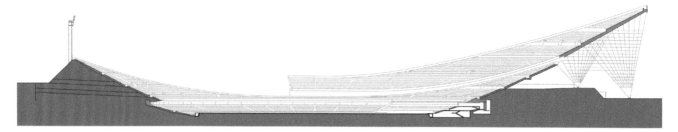

Mexico City's Estadio Olimpico was the first Summer Games venue to feature an all-weather synthetic track surface.
SOURCE: CREATIVE COMMONS – CREDIT LATOGA.

Design

To meet IOC requirements, seating capacity had to be increased substantially – to 83,700. This was accomplished without major structural modifications.

Four distinctive lighting towers were constructed, along with structures housing a new large electronic scoreboard and the Olympic Cauldron. Elevators to the presidential box and the enlarged observation "overlook" for judges and journalists were installed, facilities for the press and athletes upgraded.

From the athletes' point of view, the most significant improvement was to the track and other performance areas. There had been mounting complaints about the cinder tracks used in all previous Summer Games. The heavily trafficked inside lane was prone to deteriorate significantly over the course of the Games, especially in wet weather. Two very different athletes, Jesse Owens and Emil Zatopek, were the most high-profile complainants, indicating that both sprinters and distance runners could be handicapped by such deterioration.

In the 1960s, however, an all-weather polyurethane surfacing was developed for horse racing in the USA. Dubbed "Tartan" by its maker 3M – best known for producing Scotch Tape – the new surfacing had obvious potential as an alternative to cinders. It offered consistent

The 1968 Summer Games were highly popular with local citizens – although not all may have paid an entrance fee.
SOURCE: IOC

performance in adverse weather and improved performance over cinders overall. One weakness was the use of hazardous mercury during manufacture.

Tartan horse racing tracks never caught on, and alternative all-weather polyurethane surfacings were soon on the market, offering a mercury-free option. Nevertheless, all such surfacings were soon generically described as "Tartan", and all subsequent Summer Games and similar events have used all-weather surfacing in place of cinders.

A new drainage system completed the upgrade, transforming the venue into the first Olympic stadium of the modern age.

Legacy

Since the Summer Games the stadium has been the home of both soccer and American football teams as well as the university's sports stadium. It also hosted several games in the 1986 FIFA World Cup, although the final was held in the Estadio Azteca. Supplementary lighting towers have been added.

Notes

1. Goldblatt, David (2016) *The Games: A Global History of the Olympics*, New York: W.W. Norton, p. 262.
2. 1968: Student riots threaten Mexico Olympics. BBC Sport, 2 October. http://news.bbc.co.uk/onthisday/hi/dates/stories/october/2/newsid_3548000/3548680.stm
3. The Dead of Tlatelolco. Washington, DC: National Security Archive. https://nsarchive2.gwu.edu/NSAEBB/NSAEBB201/

CHAPTER 21
MUNICH (MUNCHEN)
WEST GERMANY, 1972

Background

At the IOC meeting in Rome in 1966 Munich topped the second round of voting for the 1972 Summer Games, beating Madrid, Montreal and Detroit. The West German post-war economic miracle was transforming the war-shattered country, and the West German government was anxious to demonstrate that it was now a modern, liberal, democratic and prosperous state.

Die Heiteren Spiele –the Cheerful or Friendly Games – was the official motto, and for the first time there was an official Olympic mascot, Waldi the dachshund.

No fewer than 11 nations made their first appearance at a Summer Games: Albania, Dahomey (now Benin), Gabon, North Korea, Lesotho, Malawi, Saudi Arabia, Somalia, Swaziland, Togo, and Upper Volta (now Burkina Faso). The IOC withdrew Rhodesia's invitation four days before the opening ceremony in response to pressure from African nations.

In all, 121 nations sent 1,059 women and 6,075 men to compete in 21 sports. Handball and archery returned to the Summer Games after many years, and slalom canoeing was held for the first time.

Although the Munich Games themselves were acknowledged to be a great success, with landmark performances in many events, they are inevitably overshadowed by the tragedy that unfolded over 18 hours on 5 September, five days before the Closing Ceremony. Eight members of the Palestinian Black September movement stormed into the Olympic Village and took nine Israeli athletes, coaches and officials hostage.

Two hostages were killed immediately. Later that day surviving hostages and their captors were helicoptered to a military airfield, where a plane was supposed to be waiting to take them to an Arab country. Instead, a rescue attempt was launched by the local police that ended in bloody failure.

All the hostages were killed, along with a policeman. Five of the hostage takers were also killed. The three captured alive were released the following month in response to the hijacking of a Lufthansa Boeing 727 in Beirut.

Stadium origins

Originally a parade ground then an airfield, the Oberwiesenfeld had been used as a dumping ground for rubble from the wartime bombing of the city, and later as an area where travelling fairgrounds could set up. Only 4km from the city centre, its 280ha offered the opportunity to create an integrated Olympic Park where the vast majority of the events could be staged.

A comprehensive terraforming exercise transformed the area into a verdant landscape of hills and hollows and a lake. The Park, and the adjacent Olympic Village, was linked to the city centre by a system of new expressways and a high-speed railway.

The design team

An international design competition was won by German structural engineer Frei Otto (1925–2015) and German architect Gunther Behnisch (1922–2010), assisted by Greek-born Stuttgart University Professor John Argyris (1913–2004).

Frei Otto became famous for his pioneering work with tensile and membrane structures from the 1950s onwards. In 2006 he was awarded the Royal Gold Medal by the UK's Royal Institute of British Architects.

One of Germany's youngest U-Boat captains in the Second World War, Gunther Behnisch was a pioneer of deconstructivism, the postmodern architectural movement. He also designed the new West German parliament building in Bonn.

John Argyris was one of the developers of the finite element method of structural analysis. He was elected as a Fellow of the Royal Society in 1986.

Design

It is impossible to consider the design of the stadium in isolation, as it is part of a linked, integrated complex incorporating the stadium, the swimming pool and the indoor arena. Despite its huge size, this was an architectural and engineering concept that was innovative and human in scale. It was "free", in total contrast to the rigid, Fascist-inspired formality of the 1936 Berlin Games: its originality signalled a break with the past, it was a symbol of the new liberal and democratic West Germany.

Like the Berlin Games, however, the main stadium was part of an integrated Olympic Park. A spectacular common roof sheltered stadium, pool and arena. The unified complex was set into and flowed into the landscaped parkland, and has been described as a "Bedouin tent draped across a fertile oasis",[1] and it was said that "the forms are full of excitement and possess all the potential energy of a drawn bow."[2]

To avoid heavy shadows on the central arena affecting colour television transmissions, a translucent roof was needed. A 200m² trial roof section was constructed to evaluate three different potential roofing materials: glass fibre reinforced polyester, polyvinyl-coated fabric and acrylic glass.

Acrylic glass panes 3m square were the final choice to make up the 75,000m² roof, suspended by a vast cable net structure. Forming the desired double curvature "saddle-shaped" cable net section over the stadium was a major challenge, met by Frei Otto's solution of so-called "air supports". These are secondary cables supported off the main roof cables.

Munich's Olympic Park has been described as a "Bedouin tent draped across a fertile oasis."
SOURCE: CREATIVE COMMONS – CREDIT TIIA MONTO

Steel pylons up to 80m high support Frei Otto's innovative cable net roof.
SOURCE: CREDIT SENA OZFILIZ

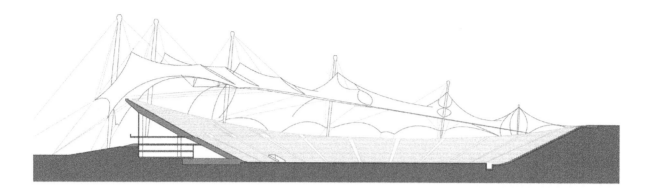

Nine continuous sections make up the roof of the stadium, supported on pylons up to 81m high. Roof edge cables were highly stressed: 80mm diameter parallel strand cables similar to those used on cable stayed bridges were selected.

There was a significant cross fall across the stadium site. On the eastern side around two-thirds of the stadium is "embedded in a hollow,"[3] with the seated and standing areas made up of 10 x 10m in situ concrete slabs anchored by 3m steel piles.

On the west side in situ concrete frames carry precast concrete box girder steps spanning up to 15m. Spectator sightlines were a major consideration: the aim was to direct "the eyes of each spectator in the stadium to a point 8.80m in front of the first row of the grandstand."[4] This resulted in a parabolically sloped layout.

Seating extends to a height of 29m above the central arena at the eastern side, while at the north and south seating has a maximum height of 14m above arena level.

Spectator capacity was 80,000, with 70 per cent of the seats covered. There was a heated turf infield area, supplied with a warm water irrigation system.

Some 70 per cent of the seating is protected by the spectacular cable net roof.
SOURCE: CREATIVE COMMONS – CREDIT TOBI 87

Legacy

Perhaps the most pervasive legacy of the 1972 Summer Games is the massively enhanced security now seen as essential for all major sporting events. The high extra costs involved were a major factor in the financial struggles of the Montreal Olympics organizing committee (see Chapter 22) and continue to be a significant factor in any bid for future Games, Summer or Winter.

Since the Olympics the stadium has been well used. It has acted as home stadium for both Bayern Munich football club and their arch-rivals TSV 1860 Munich, who shared the ground until 2005. Then they both moved to the purpose-designed Allianz stadium in the city, demonstrating once again the difficulty of combining athletics and field events in the same stadium.

Nevertheless, the stadium has hosted a number of Champion's League finals and numerous concerts by international artists. More unusual events include motorsports and, almost unbelievably, cross-country skiing. This latter was made possible by giant snow-making machines.

Outside the stadium the Olympic Park landscape and facilities continue to be a highly valued asset for the city. The giant roof has weathered well and has suffered far fewer problems than other unconventional designs.

Notes

1. Inglis, Simon (1990) *The Football Grounds of Europe*, London: Willow Books.
2. Gordon, Barclay F. (1983) *Olympic Architecture: Building for the Summer Games*, New York: John Wiley. ISBN -13: 978-0471060697.
3. Munich Official Report, Volume 2, pp. 50–54.
4. Gordon, *Olympic Architecture*.

CHAPTER 22

MONTREAL
CANADA, 1976

Background

Montreal was the rank underdog when the IOC met in Amsterdam in 1970 to decide on the host city for the 1976 Summer Games. The smart money was on one of the two superpower cities in the race, Los Angeles and Moscow. In the event representatives of the smaller countries swung the votes towards Montreal, which defeated Moscow in the second round.

Helping Montreal's case was the highly successful Expo World Fair in 1967, plus its perceived political neutrality. But it was the intensive lobbying of the smaller nations by the autocratic and controversial Jean Drapeau, the powerful mayor of Montreal, that seems to have made the real difference.

Drapeau promised delegates a "modest, self-financing Games."[1] Unfortunately, soon after the award, terrorists from the Quebec independence movement kidnapped the British Consul and a government minister, who was later murdered. Security for the Games had to be massively enhanced, at a much greater cost than anticipated.

The 1970s saw hyperinflation run riot through the global economy, and this had a catastrophic impact on Montreal's budgeting. Nevertheless, Drapeau remained resolutely positive, stating, "The Games can no more have a deficit than a man can have a baby."[2] By 1975, when it was obvious that the Games were a financial disaster, cartoons began to appear depicting a heavily pregnant Drapeau phoning Montreal's best-known abortion rights campaigner.

In early 1976 the IOC refused to ban New Zealand from the Games after the country's famous All Blacks rugby team defied UN calls for a sporting embargo on apartheid-era South Africa. Twenty-nine countries subsequently boycotted the Games, most of them African. In all, 92 nations participated. There was a total of 4,824 male competitors and 1,280 women, competing in 21 sports.

Even by the time the Games were officially opened by Queen Elizabeth II the complex and ambitious main stadium was still unfinished. Although the Montreal Games were hailed as a sporting success, Canada became the first and so far only host country not to win a single gold medal at its home Summer Games.

Stadium origins

Mayor Drapeau had long nurtured the dream of attracting a US Major League Baseball team to Montreal. For this he would need a large stadium, which, given the severity of Montreal winters, would need to be covered. Drapeau believed he could have a new stadium ready by 1971, and commissioned French architect Roger Tallibert to draw up a proposal.

Once the 1976 Summer Games had been secured it became essential for the stadium to have an opening roof, a rare feature at the time, as IOC regulations effectively ruled out a covered main stadium.

A 48ha site only 5km to the east of the city centre was the chosen location for the Olympic Park. Known as Maisonneuve Park, the area had been devoted to sports and recreation since 1912.

The architect

Roger Tallibert (b.1926) was the designer of the iconic Parc de Princes stadium in Paris, France, which opened in 1972. This featured a concrete structural frame with distinctive external "razors", echoes of which could be seen in Tallibert's design for Montreal.

There was no open design competition for any of the Olympic venues, and Drapeau faced criticism for awarding many contracts to his political cronies without competitive tendering. He is also alleged to have given Tallibert a blank cheque and urged him to create an iconic main stadium that would "brand" Montreal.

Secrecy surrounded the award of the contract to design the stadium to Tallibert, just part of the scandal that erupted during the construction phase. The scale of Tallibert's fees, reputed to total US$45m and equivalent to almost twice the entire earnings of the Quebec architectural profession in 1974, also caused a furore once revealed.

Design

Tallibert's inspiration was said to be plant and animal forms, particularly vertebral structures with sinews. He drew up plans for a futuristic complex that integrated the main stadium, the swimming pool and the velodrome – and that would push concrete technology to its 1970s

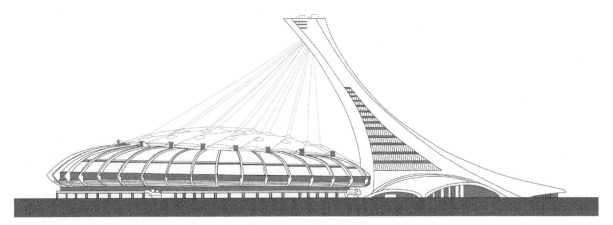

limits. It was one of the most ambitious designs of the 20th century, comparable to the hubristic 1936 Berlin Olympic Park (see Chapter 14).

Towering over the complex was to be the world's tallest (175m) inclined structure, canted at 45° and intended to suspend the Kevlar fabric opening roof via an array of steel cables. In the tower's flared base was the swimming arena, with the velodrome nestling alongside.

Long-term spectator capacity was to be 60,000, with an additional 10,000 temporary seats for the Games. Infield layout was conventional.

Post-tensioned precast concrete was Tallibert's signature structural option. Thirty-four giant "consoles" made up the main skeleton of the central elliptical "doughnut", cantilevering out up to 50m from the base, with their extremities linked together by a tubular concrete "technical ring". Each one was assembled on site from hollow precast units and tensioned together in a complex and demanding construction sequence.

All spectator seating was covered by a fixed roof, made up of a thin concrete shell lower section and a steel higher section.

Much of the secondary structure was also precast concrete. In all, around 12,000 individual precast elements were needed: most, particularly the voussoir units that made up the majority of the consoles, had to be produced to extremely demanding standards of accuracy and strength.

A brand new precast concrete plant had to be set up some 40km from the site. This took around six months to complete, due largely to the complexity of the moulds needed.

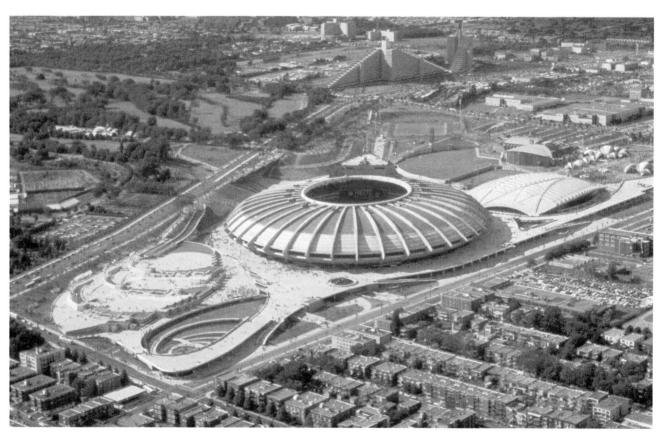

Montreal's Olympic Stadium was far from complete when the 1976 Summer Games opened.
SOURCE: IOC

Voussoirs were produced by the "match-casting" technique to ensure accuracy of assembly. Concrete was generally a rapid-hardening C50 pump mix.

Construction only began in 1973, and right from the outset it was dogged by strikes and escalating raw material costs. The largely French project management team was alleged to have made too little allowance for Canada's brutal winters, and to be short of the skills needed to supervise such a complex project. Costs soared, there were accusations of blatant profiteering on the part of the contractors and supply chain.

As it became obvious that the project was massively over budget and unlikely to be finished in time for the Summer Games the Quebec provincial government threw Tallibert and the project managers off the project. There was no hope of completing the iconic tower or installing the roof in time, but the basic stadium was ready to host the Games, which it did successfully.

Legacy

After a lengthy period in financial limbo, construction of the tower resumed in the 1980s, only to be interrupted by a large fire. A substantial lump of concrete later fell from the tower.

A local tax on cigarettes helped to finance the completion of the tower and the installation of the Kevlar fabric roof. This proved to be an ongoing disaster after it was finally completed in 1987.

It was not until 1987 that work on the tower and roof were finally completed.
SOURCE: IOC

Montreal's troubled Olympic Stadium never became the international icon Mayor Drapeau had hoped for.
CREDIT: CREATIVE COMMONS – CREDIT ALAIN CARPENTIER

It could only be opened and closed in light winds and frequently ripped. In 1998 it was replaced by a fixed fabric roof – a year later part of this roof collapsed under snow loading.[3]

Despite a multimillion dollar upgrade the roof still continued to cause serious problems, with nearly 7,000 tears recorded in 2015. A total replacement costing an estimated $250m has been mooted, and proposals for the new roof are currently (2019) under consideration. A design with removable sections is said to be most favoured.

Over the years the stadium has hosted baseball, American football and soccer teams. Its suitability for baseball was compromised by the height of the roof, which at 52m above the playing surface was well within the range of the biggest hitters, and its multisport design, which left the upper ranges of seats too far from the action.[4]

In 2018 work began to convert the 50-storey tower, previously unused, into office accommodation. Existing concrete cladding panels were to be replaced by glass curtain walling and floor slabs extended.

Originally costing $770m, by the time the final payment was made in 2006, repairs, modifications and interest paid over the three decades since the Games had inflated total cost to $1.47bn. The 1976 Summer Games has the record cost overrun of all Games, coming out at more than seven times the original estimate.[5]

Notes

1. http://montrealgazette.com/storyline/
   jack-todd-on-the-1976-montreal-olympics-drapeaus-baby-from-bid-to-billion-dollar-bill
2. www.cbc.ca/player/play/1754412484
3. https://www.newcivilengineer.com/dome-supplier-faces-montreal-compensation-battle/831802.article
4. www.espn.com/page2/s/list/worstballparks/010503.html
5. http://montrealgazette.com/sports/montreal-olympics-cost-overruns-tarnished-jean-drapeaus-legacy

CHAPTER 23

MOSCOW
USSR, 1980

Background

Moscow and Los Angeles went head to head again at the IOC meeting in Vienna in October 1974, just as they had done four years earlier in Amsterdam. This time, however, there was no rank outsider to step in and snatch the prize from under the noses of the two superpower cities (see Chapter 22). Moscow was selected by a substantial majority: Los Angeles went on to host the 1984 Summer Games, for which it was the only bidder (see Chapter 24).

These would be the first – and so far only – Summer Games to be held in Eastern Europe and the first in a nominally socialist country.

Because of its political history and its isolation from the West, the city was short of hotels and basic tourist facilities, so this was an opportunity to bolster its underdeveloped tourist industry. Russia had risen to be a superpower, and its socialist structure gave it a huge workforce, and a large standing army. This was helpful to the task of putting on an impressive event for the city.

Great resources were brought into play, even though behind the scenes some top Soviet politicians were privately appalled at the scale of the spending involved.[1] However, the desire to transform Moscow's drab, utilitarian image and show the world "an exemplary communist city"[2] won the day.

In 1979 the Soviet Union invaded Afghanistan. The move triggered widespread condemnation, particularly in the West. President Jimmy Carter announced that the US team would boycott the Moscow Games, and urged other countries to follow the American lead.

Ultimately, 65 countries failed to appear, not all in response to the US initiative. Some cited economic reasons: Iran instituted its own boycott, as it was hostile to both superpowers. Despite being one of the USA's closest allies, the United Kingdom did send a team.

In all, teams from 80 nations turned up, with a total of 1,115 women and 4,064 men competing in 21 sports.

Official drug testing cleared all competitors tested, leading to claims that these were the "cleanest" Summer Games ever. However, independent testing for the presence of testosterone using a new technique found 20 per cent with abnormally high levels.[3] This new test was subsequently adopted by the IOC.

Stadium origins

Soviet athletes competed in the Summer Games for the first time in Helsinki in 1952, winning 22 gold medals and 71 medals in total. Inspired by this success, the Soviet authorities decided to prioritize the development of athletics, and authorized the construction of a world standard sports complex close to central Moscow.

An 180ha site at Luzhniki was chosen for its proximity to the city and the ease with which it could be integrated into the transport network. The name Luzhniki derives from the former flood meadows in the bend of the Moskva River, which meant the ground was marshy. Some 10,000 piles had to be driven and 3M m³ of fill dredged from the river to create a suitable platform for the complex. What was described as "a whole area of time-worn buildings"[4] had to be demolished.

Construction began in spring 1955 and the complex opened 450 days later in summer 1956, a remarkable achievement given the severity of Moscow winters. At its centre was the main stadium, then dubbed the Grand Arena of the Central Lenin Stadium. This was later changed to the Luzhniki Stadium.

The Luzhniki Stadium was opened in 1956.
SOURCE: IOC

Design team

Original architectural design was attributed to I. Rozhin, N. Ullas and A. Khryakov "under the guidance of A. Vlasov". V. Nasonov, N. Roznikov and V. Polikarpov were credited for the structural design.

Design

A classic oval planform was adopted, measuring 301.4m on the long axis. In situ concrete was rejected as a structural option, as the technical demands of sub-zero concreting would have slowed construction in Moscow's long winter. To meet the demanding completion target the designers chose steel as the primary structural material. The steel structure supported precast concrete terracing and was clad externally with stone blocks, up to the underside of the large cornice.

Opening Ceremony for the 1980 Summer Games in Moscow.
SOURCE: IOC

Externally, the façade below the cornice was a repetition of uniform bays around the circumference. Overall, the design lacked excitement, appearing safe but rather dull. Spectator capacity was 100,000, in up to 78 rows of seats, with only the upper rows sheltered by a 10m wide roof.

Behind the façade was extensive accommodation, which was the focus of much of the refurbishment and renovation carried out in preparation for the Summer Games.

This major upgrade was launched soon after the city was awarded the Summer Games. Grandstand structures were reinforced, and what the Official Report describes as "a new wet seal of cast-in-situ self-stressing concrete" was placed over the area under the grandstands – presumably to protect against seepage from the marshy ground below.

A new synthetic surfacing was applied to the eight-lane 400m running track. All new plastic seating was installed. The Olympic Flame Bowl was erected above the eastern grandstand, and there was widespread upgrading of the internal facilities. On the second floor were no fewer than 14 large gymnasiums, plus restaurants and a cinema, and on the third there was a hotel for 360 guests.

The most distinctive transformation was the erection of four 86m tall lighting towers outside the stadium proper. These could achieve a lighting intensity on the infield of more than 1,500 lux, enough to ensure high-quality colour television broadcasting.

The 86m tall lighting towers ensured high-quality colour TV broadcasts.
SOURCE: CREATIVE COMMONS – CREDIT SERGEY GUNEEV

During the Games, transport facilities included metro, bus and trolleybus, and taxis. Tourists were brought to the stadium by special buses, cars and riverboats.

Legacy

One long-lasting benefit of the construction of the original sports complex was the creation of a tree-filled park landscape around the facilities, something much appreciated by Muscovites.

Before the Games the stadium was used intensively for many sports and championships, beginning with the first Soviet Spartakiada[5] in 1956. Post-Olympics the stadium continued to be used frequently, despite the 1982 tragedy in which a crowd stampede during a UEFA Cup match killed 66, mostly adolescents.

A major refurbishment in 1996 saw a free-standing roof added and new seating installed. This reduced capacity to 78,000. An artificial grass pitch was laid in 2002.

After nearly six decades a near total reconstruction was planned, beginning in 2013. Everything internal was marked for demolition and replacement. The outer façade, however, was retained, as was the free-standing roof. The stadium is now a dedicated field sports venue, with no running track and spectators much closer to the football pitch.

Completion was in 2017, in time for the 2018 World Cup final. Thus the stadium joins those of Rome, Berlin, Munich and London as the only Olympic Stadia to also hold this event.

The Luzhniki Stadium after its 2017 refurbishment.
SOURCE: CREATIVE COMMONS – CREDIT MOS.RU

Notes

1. https://themoscowtimes.com/articles/a-tale-of-two-olympic-cities-moscow-1980-and-sochi-2014-30036
2. J. Parks (2009) "Red Sport, Red Tape: The Olympic Games, the Soviet Sports Bureaucracy and the Cold War 1952–1980". UNC Chapel Hill doctoral thesis.
3. https://www.economist.com/blogs/graphicdetail/2016/07/daily-chart-15
4. History of the Luzhniki Sports Complex. http://eng.luzhniki.ru/content/about/history/complex.php
5. Soviet Spartakiads were internal Soviet Union events comprising both summer and winter games. Extraordinary competitor numbers were claimed.

CHAPTER 24

LOS ANGELES
USA, 1984

Background

When it met in Athens in 1978 the IOC was aware that it was facing a crucial moment in the history of the Games. Montreal 1976 had been a financial disaster; Moscow's plans for the 1980 Games were equally ambitious – and expensive. It is now known that there were those at the highest levels of the Soviet government who were beginning to have second thoughts about the wisdom of spending so much money on the Games when the Soviet economy was already under severe strain.[1] And there were very few potential host cities prepared to bid for the 1984 Games or to commit to the levels of expenditure typified by Montreal and Moscow.

Tehran did express a preliminary interest, but this came to nothing following the political upheavals in Iran that culminated in the fall of the Shah in 1979. The exception was the United States. Two American cities were seriously interested, but, as only one bid per country was allowed, both had to lobby the United States Olympic Committee (USOC) for selection as the approved US bidder.

As the winner would be the sole bidder for the 1984 Summer Games the USOC vote would effectively select the host city. In the event, Los Angeles triumphed over New York. And as the sole bidder, Los Angeles was able to extract various concessions from the IOC; not least that USOC would not be responsible for any cost overruns.

There would be maximum use of existing venues, with only a new aquatics stadium and velodrome to be constructed, and these would largely be funded by corporate sponsors. Such sponsorships were a major element in the overall funding of the Games.

A further boost came from the contracts for television rights. Los Angeles received twice the amount paid for the rights to the Moscow Games and four times that for Montreal (adjusted for inflation). This was the first time that the revenue from television played a significant role in the overall financial strategy for a Summer Games, and it was particularly welcome as the contracts were signed early in the planning process.

For the third Summer Games in succession, a politically motivated boycott hit competitor numbers. Following the anti-apartheid boycott by African countries at Montreal (see Chapter 22) and the US-led boycott of the Moscow Games over the Soviet invasion of Afghanistan (see Chapter 23), 14 Eastern bloc countries refused to attend, a move that was seen as a "tit-for-tat" response to the Moscow boycott. Libya and Iran also stayed away, but for different reasons.

Nevertheless, 140 nations were represented by 6,829 athletes, 1,566 of them women. There were 221 events in 21 sports.

These numbers were boosted by the attendance of a team from the People's Republic of China for the first time since 1952, following an IOC decision to reclassify Taiwan as Chinese Taipei rather than the Republic of China.

In 1932 and earlier at the St Louis Summer Games in 1904 the then US presidents had been conspicuous by their absence. This time the Games were opened by President Ronald Reagan, a former Governor of California.

Origins

Over the five decades since it hosted the 1932 Summer Games, the Los Angeles Memorial Coliseum had been gradually upgraded. The major change was the switch from the original very basic bench bleachers[2] to individual theatre-style seats in 1964, reducing capacity to 92,516.

New electronic scoreboards appeared, along with a much enlarged press box and travertine marble cladding to the iconic concrete peristyle. Three escalators and 54 lavatories were installed, much improving spectator comfort. Spectator capacity was reduced again, to 71,500, when a larger eastern grandstand was opened in 1978.

From the point of view of the football and baseball players, the biggest breakthrough was possibly the long-awaited arrival of air conditioning in the locker rooms in 1983. In all, there were 90 entrances to the Coliseum and 74 turnstiles.

For more information on the origins of the Coliseum, see Chapter 13.

First constructed in 1932, the main in situ concrete structure is still in good condition.
SOURCE: CREATIVE COMMONS – CREDIT MIKEJIROCH

Design team

Original architects were the father and son team of John B. Parkinson and Donald D. Parkinson.
Latest upgrades are the responsibility of the multidisciplinary DLR Group.

Design

Despite its gradual evolution since the 1932 Summer Games, the Coliseum was still short of the standards expected by the IOC. One fundamental nonconformity was the layout of the vital 400m running track. The minimum radius of the curves on the existing track was tighter than the IOC's specified minimum of 36.5m – but an acceptable layout with larger radius curves would not fit into the existing infield.

So the decision was taken to demolish the first row of seats on the north side of the stadium, which created just enough space to squeeze in the new eight-lane track. The original track had been resurfaced in 1972 using a synthetic Tartan material first seen at the 1968 Summer Games in Mexico (see Chapter 20): this time around the similar Rekortan surfacing was specified. Rekortan was first used at the Munich Summer Games in 1972.

It was also considered essential to improve the infield, partly with its planned subsequent

Performers and spectators await the lighting of the Olympic Flame at the 1984 Summer Games in Los Angeles.
SOURCE: CREATIVE COMMONS – AUTHOR UNKNOWN

use as an American football field in mind. Drainage was improved, and a high performance grass variety planted.

A major upgrade of the Coliseum's sewerage system was also deemed essential. Electric power supplies were significantly boosted, and more powerful air conditioning was installed in the locker room.

Two massive new electronic scoreboards were planned, but it soon became obvious that the existing concrete structure would be unable to support their weight. Large I-beams had to be inserted into the peristyle arch complex to support the scoreboards, and the peristyle refurbished.

With the increasing importance of TV coverage the light levels in the stadium came under scrutiny. All points on the track and infield had to achieve 200 foot-candles (2,153 lux) of illuminance and at first it was thought that replacing the existing 348 lamps with 1,500-watt metal halide lamps would be the answer.

However, it was soon discovered that in practice the ends of the track fell short. Additional temporary light towers were installed at each end of the stadium in response.

There were many other minor improvements, upgrades and refurbishments, not least the installation of a new public address system. Much of this was funded by sponsors, with the organizers contributing over $5m.

This bronze statue by Robert Graham was installed outside the Memorial Coliseum ahead of the 1984 Summer Games.

Despite the use of temporary seating, spectators could be a very long way from the action during American football games.
SOURCE: CREATIVE COMMONS – CREDIT BOBAK

Legacy

The Coliseum made history in 1984 when it became the first stadium ever to host the Summer Games twice. Also in 1984 the stadium was declared a National Historic Landmark, confirming its official recognition as a structure of outstanding historical significance.[3]

Throughout its life the Coliseum's main function was to host American football and baseball games, for which it proved to be not well suited. Its main drawback was its size. As a baseball venue the Coliseum's infield was too small and the wrong shape, while football spectators could be a very long way from the action.[4]

In the early 1990s a radical remodelling took place. The infield was lowered by 3.4m and 14 new rows of seats replaced the running track. There were plans for an even more radical transformation – then the 1994 Northridge Earthquake caused serious damage to the stadium.

Nearly $100m had to be spent on repairs, and the radical plans were abandoned. In 2015 the USC unveiled a $270m plan to renovate and restore the Coliseum and create new amenities for spectators.

Following the award of the 2028 Summer Games to Los Angeles it was announced that the Opening and Closing Ceremonies would be held in the Los Angeles Stadium at Hollywood Park, due to open in 2020. The soccer competition would also be held there. Athletics, however, would return to the Coliseum for a record third time, which will require the installation of a new running track.

Notes

1. https://themoscowtimes.com/articles/a-tale-of-two-olympic-cities-moscow-1980-and-sochi-2014-30036
2. Bleachers are uncovered terraces of seats, often without backrests. Traditionally constructed of wood with sun-bleached plank seating, usually open to the ground below, bleachers are now seen as low status accommodation.
3. https://npgallery.nps.gov/pdfhost/docs/NHLS/Text/84003866.pdf
4. Schwarz, Alan (2008) "201 feet to left, 440 feet to right: Dodgers play the Coliseum", *The New York Times*, 26 March.

CHAPTER 25

SEOUL
SOUTH KOREA, 1988

Background

Only two cities bid for the 1988 Summer Games, both of them Asian. At the IOC meeting in Wiesbaden, West Germany, in 1981, Seoul easily defeated the Japanese city of Nagoya. This decision was to prove momentous for the young country, which came into existence only in 1948 after the defeat of Japan in the Second World War. The Korean peninsula was split into two, with a communist North backed by China and the Soviet Union, and the South, supported by the United States.

In 1950 North Korea invaded the South. Three years of bitter fighting ensued. Local forces were eventually bolstered by troops and military assistance from several Western nations in the case of the South and from both China and the Soviet Union in the case of the North. Seoul changed hands four times between the North and South Korean forces, and was heavily damaged: it was later estimated that more than 190,000 buildings, 55,000 homes and 1,000 factories were destroyed.[1]

After an armistice was signed in July 1953 South Korea was led by the authoritarian Syngman Ree. He was finally overthrown in 1960, but the country soon fell under military rule, with supreme authority vested in Major General Park Chung Hee.

Park launched a massively successful industrialization programme and was considering an Olympic bid when he was assassinated by the head of the Korean Central Intelligence Agency in 1979. His successor, another general, Chun Doo-hwan, revived the bid in 1981.

By this time the increasingly repressive military dictatorship was meeting serious resistance from the burgeoning urban middle class. There were widespread protests and riots, which were put down with extreme brutality. It began to look as though Seoul would lose the Games. Then, after a secretive meeting with IOC President Juan Samaranch, Chun backed down, and granted sweeping political freedoms.[2]

The South Korean government was also briefly pressured by the IOC to collaborate with the North in staging a joint Games. The North's demands and conditions were very soon declared to be unreasonable and unacceptable, and the attempt failed.[3]

Tokyo 1964 was seen as the model by the military junta. They hoped that a successful Games would similarly mark South Korea's entry into the family of First World nations, while at the same time legitimizing their autocratic regime.

At the time Seoul had a population of 10M, while there were 15M people living in surrounding areas. Generous funds were made available; the city received a major cosmetic overhaul, along with upgraded telecommunications and transport infrastructure.

In a typically brutal move, the authorities rounded up homeless inhabitants of Seoul and incarcerated them in notorious "work camps", where torture was rife and many died.[4]

North Korea inevitably boycotted the Games, and was joined by Cuba, Nicaragua, Ethiopia, Albania and the Seychelles. A team from Madagascar was expected, but never turned up. Nevertheless, a record 159 nations sent a total of 2,914 women and 6,917 men to compete in 29 sports.

These were to be the last Summer Games for the Soviet Union and East Germany: both states would cease to exist before the next Games in Barcelona.

One unusual feature was the staging of major events, particularly athletics finals, very early in the morning. This was to ensure they could be watched live on US TV at peak viewing times.

Stadium origins

Plans for a vast new sports centre 11km to the south-east of the city centre were first hatched in the late 1960s. This was intended to host the 6th Asian Games in 1970, but even after South Korea pulled out of the event on security and financial grounds the project continued.

Apart from a 100,000-seat main stadium, the complex ultimately housed a number of other sports buildings, including the large Jamsil Gymnasium, the swimming pool and a baseball stadium. There was also a warm-up pitch and track on the site.

The site is bounded by the Han and Tanchon Rivers. Their beds are nearly dry, except in the monsoon period, so for much of the year they can be used as parking lots. New subway stations were placed to the south of the complex.

Ground was first broken in 1976 and the complex was completed in time to host the 10th Asian Games in 1986.

Design team

Kim Swoo-geun (1931–1986) was a prominent local architect once described by *Time* magazine as the "Lorenzo de Medici of Seoul." Among his most well-known projects are the country's National Science Museum, the US Embassy in Seoul and Korea's Expo 70 pavilion. The architectural practice he founded in 1961 evolved into the SPACE Group, now a multidisciplinary international enterprise. Kim was responsible for the architectural design of the entire sports complex, including the main stadium. Structural engineering was undertaken by a collaboration between Seoul Architects and the Structural Engineers Association Institute.

Design

Locating the main stadium towards the Han River bank to the north-west meant that all 100,000 spectators could reach mass transit stations within 10 minutes of an event closing.

Seoul's Olympic Stadium is surrounded by other Olympic venues.
SOURCE: IOC

Kim adopted a classic circular format for the exterior shape. According to the Official Report, his inspiration for the profile was a Korean Joseon Dynasty porcelain vase. Inside, the stadium has more seats distributed to the sides, which are consequently higher. This is the preferred arrangement for spectators. The curving roof follows this configuration.

Spectator access is via an upper large podium or concourse around most of the stadium. Athletes, officials and media use the access at the lower level.

In situ reinforced concrete was the material of choice for the spectator areas. A ring of 80 curved concrete columns or ribs around the exterior gave it a consistent rhythm and brought unity to the design.

These columns or ribs support the cantilevered steel roof beams, which vary between 30m and 40m in length. A proportion of the seating was left open and uncovered.

Whilst the stadium does not have the kind of exciting structure found in some Olympic stadia, this was a very competent, unified and convincing design, which encompassed the best

The architect is said to have been inspired by classic Korean ceramics.
SOURCE: CREATIVE COMMONS – CREDIT CHELSEA MARIE HICKS

Seoul's spectacular Olympic Cauldron was the focus of the Opening Ceremony of the 1988 Summer Games.
SOURCE: IOC

design practice. Nevertheless, it provided a significant construction challenge, with the project taking nearly seven years to complete.

Each of the 80 supporting columns was unique in height, cross-section and curvature. More than 160,000m^3 of ready mixed concrete was needed, and a 222t crane had to be imported to erect the roof beams, which weighed up to 36t. A specially developed "automatic pulley" was needed to persuade the roof canopy to conform to the sweeping curves of Kim's design.

Legacy

Seoul 1988 is perhaps best remembered for an infamous event involving the Canadian sprinter Ben Johnson, who set a world record in the 100m final, only to be disqualified for the use of performance-enhancing drugs.

The stadium itself hosted the national football team until 2000, then was primarily a concert venue until 2015, when it became the home of the newly formed professional football club Seoul E-Land FC.

Notes

1. Hamnett, Stephen and Dean Forbes (eds) (2012) *Planning Asian Cities: Risks and Resilience*, Abingdon: Routledge. p. 159.
2. Cho, Ji-Hyun and Alan Bairner (2011) "The sociocultural legacy of the 1988 Seoul Olympic Games", *Leisure Studies*, 31(3): 271–289. doi:10.1080/02614367.2011.636178
3. Radchenko, Sergey (2011) "Sport and politics on the Korean peninsula – North Korea and the 1988 Seoul Olympics", NKIDP e-Dossier No. 3.
4. http://boingboing.net/2016/04/20/pre-1988-olympics-south-korea.html

CHAPTER 26

BARCELONA
SPAIN, 1992

Background

Barcelona had bid for the 1936 Summer Games but lost out to Berlin. In 1986, however, the city scored a third-round victory over Paris, Brisbane, Belgrade, Birmingham and Amsterdam in the bidding to host the 1992 Games. The event would be the showcase for the decade-long transformation of the city, from a decaying post-industrial Mediterranean seaport into a modern cultural centre, a high-profile tourist destination and an architectural jewel.

Barcelona's city council had launched its campaign for the Summer Games in 1981, seeing the event as a way of attracting the funding it needed to complete its ambitious plans for urban renewal. It was backed by the regional Catalan government, and by the Spanish government, which was eager to promote a more modern, high-tech image of the country.

Such support was vital. The 1992 Games would be the most expensive since Tokyo 1964.[1] With the funding on tap, the city was now able to complete a number of town planning works that had been postponed for many years. The recovery of the beaches and the sea front, the construction of the ring roads and metro system, the upgrade of the airport and sewerage systems and the large-scale telecommunications works were all carried out as a result of the Games.

The overall planning strategy finally adopted was to create four large Olympic areas, a continuation of the tradition of Olympic Parks of previous Games. In the case of Barcelona, these areas were an integral part of the urban fabric of the city. In them were concentrated more than 80 per cent of the competition facilities. They were to offer a wide range of sports and leisure facilities with excellent access from within the city and to the wider Catalonia.

There was a search for variety and quality in the architectural solutions, which resulted in the appointment of internationally renowned designers following an international design competition. The resulting prestigious architecture was a bonus for the city and avoided the monumentality of its predecessors in Montreal and Seoul.

As a result of the geopolitical turmoil following the break-up of the Soviet Union and the collapse of Yugoslavia there were many new nations represented at these Summer Games. Twelve of the former Soviet states sent a Unified Team: for the first time since 1936 the Baltic states of Latvia, Lithuania and Estonia competed as independent nations. Croatia, Bosnia-Herzegovina and Slovenia had their own teams, and Yemen and Namibia also made their debuts.

South Africa reappeared after a 32-year absence after the end of apartheid. Yugoslav athletes had to compete as Independent Olympic Participants due to UN sanctions against the Federal Republic of Yugoslavia.

In all, 169 nations sent 2,704 women and 6,652 men to compete in 25 sports. For the first time since the earliest days of the Summer Games professional athletes were allowed to participate. The US "Dream Team" of basketball players from the professional National Basketball Association league easily defeated Croatia in the final.

This was the last time that the Summer and Winter Games were staged in the same year.

Stadium origins

Located in the Montjuic Olympic park on a hill to the south-west of the city centre, what was originally known as the Estadi de Montjuic was opened in 1929 after a two-year construction phase. Its immediate role was to host the city's 1929 International Exposition – its secondary function was to boost Barcelona's unsuccessful bid for the 1936 Summer Games.

After the Exposition the stadium was little used, serving mainly as an athletics training ground. Post-Second World War it hosted the occasional soccer match and acted as the paddock for the nearby Montjuic racing circuit. In the 1970s it began to deteriorate and was virtually abandoned except by the city's homeless.

Design team

Architect Pere Domènech I Roura was responsible for the 1927 design. For the Olympic renovation a multinational team of architects was assembled, featuring Italian architect Vittorio Gregotti and Catalan architects Federico Correa, Alfons Milà, Joan Margarit and Carles Buxadé.

Design

Although the interior of the stadium was judged to be beyond economic repair the elaborate neoclassical stone façade was in reasonable condition and was to be retained. This restricted the architects to the unusual original planform, with long straight sides linked by large radius curves, and also limited the overall height of the seating tiers.

There was a pressing need to increase spectator capacity to meet IOC standards. In response the design team chose to rip out the old decaying stands, excavate down 11m at arena level and build up new grandstands that gave a final capacity of 60,000.

Originally part of the seating on the western side was protected by a steel roof, by now much decayed. The new 150m x 30m tensile fabric cantilever roof was supported by a 6.3m high triangulated tubular steel truss spanning between three reinforced concrete columns, with the backspan tied down by 32 tensioners.

A curvaceous steel fabrication was bolted to the outer façade of the stadium at the northern end to support the Olympic Cauldron. A synthetic 400m nine-lane running track was laid,

Retaining the original 1929 stone façade presented the designers of the upgraded stadium for the 1992 Summer Games with some unusual challenges.
SOURCE: IOC

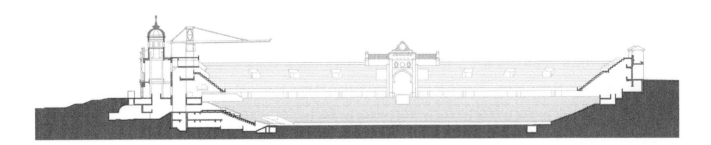

A curvaceous steel fabrication was simply bolted to the stone façade to support the Olympic Cauldron.
SOURCE: CREATIVE COMMONS – CREDIT BALOU46

surfaced with a Mondo-Enimont polymer. Modern communication and service facilities were installed, and lighting towers erected. There was a 60m warm-up track below the main grandstand, along with many service passageways.

Legacy

Perhaps the most influential legacy of the 1992 Summer Games was the opening of the door to professional athletes. By this time the Games were increasingly dominated by what were known as "shamateurs", i.e. those athletes who were covertly subsidized by governments or academic institutions to train and practise full time. This recognition of the new reality moved the Olympics into a new era.

In 1989 the stadium was renamed Estadi Olímpic de Montjuic. This was changed again in 2001 to Estadi Olímpic Lluís Companys, to honour a former Catalan president who was executed by the Franco regime in 1940.

Since the Games the stadium has struggled to find a long-term role. There have been two failed attempts to base professional teams there: First Division football club R.C.D Espanyol and a local American football club both failed to put down roots, the latter suffering from

The neoclassical stone façade has weathered well.
SOURCE: CREATIVE COMMONS – CREDIT CANAAN

the relative shortness of the playing area. This was eventually lengthened to accommodate a regulation size, but there have been only a few matches played on it.

In 2010 the stadium hosted the European Athletics Championships and in 2012 the Junior World Athletics Championships was held there. It is now a stop on the tourist circuit in Barcelona.

Otherwise the legacy for the city has been positive. Barcelona is now seen as a very desirable place to live and visit. By the end of the 1990s it had become one of Europe's most visited places, after Paris, London and Rome, to the point where mass tourism is causing friction with locals at peak times.

Note
1. Flyvbjerg, Bent and Allison Stewart (2012) "Olympic proportions: cost and cost overrun at the Olympics 1960–2012", Working Paper. Saïd Business School, University of Oxford.

CHAPTER 27

ATLANTA
USA, 1996

Background

Atlanta was originally seen as the rank outsider when six cities declared their bids in 1990. Toronto was the front-runner, Canada having staged a successful Winter Olympics in 1988, and there was a strong argument that as 1996 would be the centenary of the first modern Games in Athens it would be fitting for it to host the Games again.

The United States had already hosted the Summer Games on four occasions, the last only 12 years earlier in Los Angeles. However, Atlanta mounted the largest public relations and lobbying operation in Olympic history, making much of its image as the birthplace of the civil rights movement in the USA. At the 1990 IOC meeting it eventually defeated Athens in the fifth round of voting. Toronto only managed third place, Melbourne, Manchester and Belgrade were also-rans.[1]

Los Angeles 1984 was seen as the model for the Atlanta Games, with much of the finance coming from sponsorship and TV rights. In reality, massive injections of public money were needed to finance the upgrade of the airport and many essential infrastructure improvements.

As at a number of previous and subsequent Summer Games, there were sustained campaigns to clear the streets of more than 9,000 homeless people[2] and to demolish poorer, largely African-American, residential areas to free up land for new Olympic infrastructure.

There was no investment in new urban transport; instead, a large fleet of buses was drafted in from all over the state of Georgia. These were not a great success, as their out-of-town drivers were unfamiliar with the city and frequently got lost.

During the Games a pipe bomb exploded in the Olympic Park, killing one spectator and triggering a fatal heart attack in a Turkish cameraman. It later transpired that the bomber was protesting against Federal government policy on gay rights and abortion on demand.[3]

One of the most poignant moments of the Games was the lighting of the Olympic Flame by the visibly ageing Mohammed Ali. In all, 10,318 athletes competed in Atlanta, 3,512 women and 6,806 men. There were no boycotts, so a record 197 nations were represented.

There was controversy at the Closing Ceremony when IOC president Juan Samaranch failed to hail the Atlanta Games as "the best Olympics ever," as was the tradition Samaranch himself had inaugurated. This was seen as a protest at the overtly commercial ethos of the event.

Stadium origins

Right from the outset the event organizers planned to transform the new stadium into a baseball venue after the Games. This was the first time that a specific Olympic legacy use was predetermined at the design stage, posing some unique challenges for the design team. Earlier attempts to convert the Los Angeles Memorial Coliseum into both an American football field and a baseball park had been less than successful (see Chapter 13).

The same applied to the nearby Atlanta-Fulton County Stadium, which had only been opened in 1965. It was shared by the major league Atlanta Braves baseball team and the American footballers of the Atlanta Falcons, and proved satisfactory for neither. In 1991 the Falcons moved to a new covered football stadium, the Georgia Dome, and the fate of the County Stadium was sealed.

Design team

Four multidisciplinary design practices came together in a joint venture to take on the stadium project. Lead management was provided by Heery International, supported by Rosser International, Williams-Russell and Johnson, and Ellerbe Becket. Heery also served as prime client contact.

Design

A distinctively asymmetric layout was the design team's response to the overriding requirements of the legacy function. Precast concrete construction was adopted to simplify the post-Olympic conversion. Specified spectator capacity was 85,000 during the Games, reducing to 45,000 for the eventual ballpark.

An early rendering showing the planned design with provisions for after-use already included.

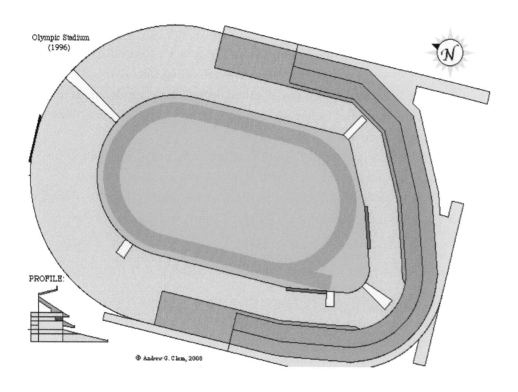

Olympic Stadium
(1996)

PROFILE:

© Andrew G. Clem, 2008

Designed for a specific legacy use as a baseball stadium, Atlanta's Centennial Olympic Stadium had a unique configuration.

Two lower tiers of seating ran round the entire stadium. The southern half was topped with two further tiers, which also contained the luxury boxes and facilities for press and broadcast media.

Seats at the south-western corner of the stands were somewhat further away from the Olympic action than was the norm for modern stadia. Lines of sight, however, were protected by the creation of a "photographers' moat" around the entire infield.

The increasing impact of the Opening and Closing Ceremonies was marked by the inclusion of a 2.7m diameter tunnel under the infield, emerging at its centre. This was to be used for special effects: according to the Official Report "adding this access, as well as all other utilities required, altered the original construction plans considerably."

For the Opening Ceremony the planning team envisaged the athletes entering the stadium down a large ramp, which would also be used for choreographed displays by professional dancers. Several hundred seats and their supporting structure had to be removed to accommodate the ramp, and then reinstalled before the start of the athletics events seven days later.

Another distinctive feature was the dramatic, free-standing Olympic Cauldron adjacent to the stadium proper. Located to the north, this stood tall enough to be visible to all the spectators inside the stadium.

Such was the quality of construction that the project received the 1996 Award of Excellence for Public Works from the Georgia Chapter of the American Concrete Institute.

Legacy

Reconfiguration began immediately the Paralympic Games closed at the end of August. The running track was removed intact, modified and gifted to Clark Atlanta University; sections of the northern end seating were dismantled, reducing spectator capacity to 47,000. This marked the end of what was formally known as the Centennial Olympic Stadium, making it the shortest lived Olympic stadium in history.

Once the conversion was complete and the Braves had moved into their new stadium, now renamed Turner Field, the

Atlanta's Olympic Cauldron had a distinctive industrial supporting structure.
SOURCE: CREATIVE COMMONS – COURTESY OF AUSTRALIAN PARALYMPIC COMMITTEE

County Stadium was demolished and its site became a parking lot. The Olympic Cauldron, however, survived.

As a major league ballpark the remodelled stadium functioned well.[4] However, its owners soon realized its city centre location was far from ideal.

Traffic congestion and limited car parking space discouraged fans, and the venue was some distance from the nearest public transport. It was also close to some of the less affluent areas of the city.

So the Braves decided to move to a new stadium on the edge of the city, playing their last season at Turner Field in 2016. Demolition loomed, but in October 2016 it was announced that Turner Field would become the new home of the Georgia State University football team. In 2017 it was renamed the Georgia State Stadium.

Notes

1.   www.georgiaencyclopedia.org/articles/sports-outdoor-recreation/olympic-games-1996
2.   Flowers, Benjamin S. (2017) *Sport and Architecture*, Abingdon: Routledge.
3.   www.findingdulcinea.com/news/on-this-day/July-August-08/On-this-Day--Bomb-Explodes-in-Atlanta-s-Olympic-Park.html
4.   https://archive.nytimes.com/www.nytimes.com/specials/olympics/0730/oly-stadium-braves.html

CHAPTER 28

SYDNEY
AUSTRALIA, 2000

Background

There were five cities in the race to host the 2000 Summer Games when the IOC met in Monaco in 1993. In the early rounds Beijing made the running, with Sydney close behind and Manchester, Berlin and Istanbul trailing. The fourth and final round was a two-horse race. Sydney came from behind to snatch victory from Beijing by just two votes, 45 to 43.[1]

This was to be the second Summer Games in the Southern Hemisphere, following on from Melbourne in 1956. Unlike Melbourne, however, there would be equestrian events, made possible by specially revised quarantine regulations. Also unlike Melbourne was the lack of high level political controversy after the award of the Games.

Nevertheless, there were protests, most notably by the more radical fringes of the Aboriginal population, but also including those objecting to the impact of Games-linked development on local rents. As the Opening Ceremony grew closer, however, these protests fizzled out and the Games enjoyed widespread public goodwill.

In 1956 12 nations boycotted the Melbourne Games, for a variety of reasons (see Chapter 17). No such boycotts marred the Sydney event. In all, 199 nations were represented, sending 4,069 women and 6,582 men to compete in 28 sports.

Afterwards there was general agreement that Sydney had taken the Olympic Games to new levels, and that its success would be hard to surpass. Comments such as "any city considering bidding for future Olympics must be wondering how it can reach the standards set by Sydney"[2] were typical of the international press reaction.

Stadium origins

One of the main reasons Sydney triumphed in Monaco was its bid's emphasis on environmental issues – this was to be the "greenest" Games so far. The site chosen for the new main stadium and other facilities typified this approach.

Homebush Bay was 760ha of contaminated post-industrial land close to the heart of Sydney. Once extensive tidal wetlands and thick bush, and an important meeting place for Aborigines, the site had been utilized for farming, a saltworks, an abattoir, brickworks, arms depot, asylum, school and prison, and a racecourse.

To allow these developments to take place the low-lying mangrove swamps had been progressively reclaimed and filled with a wide variety of materials. By the 1990s it was estimated there were 9M.m³ of domestic, commercial and industrial waste on the site.

The clean-up operation was the largest ever in Australia and one of the largest remediation projects anywhere in the world. Apart from dealing with the contaminated land and industrial waste, it also involved the restoration of the main waterway through Homebush Bay, transforming it from a heavily silted concrete channel back to its original course. More than 1.4M.m³ of mud and debris had to be removed.

Another major challenge was the presence of 47 transmission towers supporting 10km of overhead high voltage power lines. These were replaced with underground cables in an expensive operation partly funded by sponsorship.

At the outset it was decided to use the city's heavy rail system as the primary public transport to the site, and to ban the use of private transport. A new 5.3km rail loop was constructed, with 1km underground leading to a new below-ground station just 400m from the main stadium. Trains at two-minute intervals could transport up to 50,000 people an hour in and out of the Bay.

Design Team

Architectural design was carried out by Bligh Lobb Architecture, a team brought together for the Games. Lobb has now become Populous. Structural design was the responsibility of Sinclair Knight Merz (SKM) and Modus.

Design

Completed in 1999, the stadium was the largest Olympic stadium ever built up to that time. During the Games spectator capacity was 110,000, thanks to extra temporary seating at the ends, which was later removed.

The philosophy behind the design was to produce a multifunctional and economically viable venue – with impeccable green credentials. To that end, the design team created what was not just the largest but also the most technologically advanced Olympic stadium ever built until then.

Rainwater was recycled for flushing toilets. The need for air conditioning was minimized through ingenious designing for natural ventilation. Natural cooling and heating and the use of daylighting reduced energy consumption. Toxic and polluting materials were avoided.

Soaring 58m above the stadium is a hyperbolic paraboloid roof, often described as "saddle shaped".[3] This slopes down towards the infield, enhancing the usually intense spectator atmosphere and optimizing acoustic performance.

Translucent polycarbonate sheeting was chosen as the roofing material to even out the contrasts in lighting and minimize the shadows cast on to the infield. The 30,000m² of polycarbonate are supported by a remarkably lightweight steel lattice roof structure, which in its turn sits on the seating structure, and two 295m long steel trusses.

Sydney's Stadium Australia was the largest, most technologically advanced and most sustainable Olympic stadium ever built.
SOURCE: ALAMY

A lightweight long span steel truss supports a polycarbonate roof.
SOURCE: CREATIVE COMMONS – CREDIT BRIAN PRACY

Spiral ramps at each corner of the stadium optimized spectator movement.
SOURCE: ALAMY

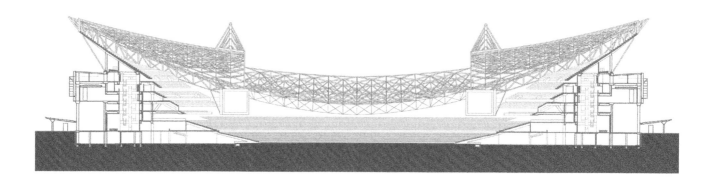

Four spiral ramps at the corners cope with moving large numbers of people. These lead into the open spaces surrounding the stadium, allowing dispersal to the nearby train station, the buses and the ferries. Lifts and escalators supplement the movement of spectators.

The stadium has been configured to accommodate future digital technology. Each seat can be fitted with small screens providing both replays and a range of information.

In a radio broadcast, IOC President Juan Antonio Samaranch stated: "What can I say – it is the best Olympic stadium I have seen in my life."

Legacy

Now best known as Stadium Australia (although its official title is the ANZ Stadium) the stadium has proved to be a very popular venue for both sporting events and entertainment. As soon as the Games and Paralympic Games were over the temporary seating was removed, reducing spectator capacity to 80,000.

In 2001 a reconfiguration was begun to accommodate sports such as cricket and Australian rules football that required an oval field. The athletics track was removed and replaced with movable seating, and awnings were added over the two ends. Capacity in the oval format was 82,500, while a rectangular set-up could seat 84,000.[4]

Rugby League, Rugby Union and soccer have all been played there over the years, along with cricket and Australian Football League games. Many cup finals and international matches have taken place. The stadium has also hosted many concerts, and both motorcycle speedway and American gridiron football have appeared occasionally.

Like several other former Olympic stadia, the venue has struggled to reconcile the needs of spectators for different sports. Spectators in the rounded ends complain they are too far away from the action when matches are taking place on the rectangular pitch.

In 2015 the New South Wales government announced controversial plans to upgrade the stadium. This would involve converting it into a permanent 75,000-seat rectangular stadium, with retractable seating at the northern and southern ends, a complete redesign of the entire stadium bowl, and a retractable roof. These plans were heavily criticized, as they would lock out both cricket and Australian rules football.

Two years later the same government decided to demolish the stadium and replace it with a new 75,000-seat rectangular stadium. The following year the government changed its mind again. Stadium Australia will now be refurbished and reconfigured as a rectangular stadium, as in the 2015 plans.

Notes

1. IOC Vote History.
2. Mossop, James (2000) "Sydney has set the highest standards for future hosts", *The Daily Telegraph*, 1 October.
3. John, Geraint, Rod Sheard and Ben Vickery (2013) *Stadia: The Populous Design and Development Guide*, Fifth Edition, Abingdon: Routledge.
4. Alan Patching & Associates, *Stadium Australia – Redefining the Customer in Stadium Design and Construction*.

CHAPTER 29

ATHENS
GREECE, 2004

Background

In 1990 Athens had felt it was entitled to stage the 1996 Summer Games, given that it had hosted the very first modern Games one hundred years earlier. IOC delegates meeting in Tokyo ultimately disagreed, judging the bid to be arrogant and lacking in detail.[1] After leading in the first four rounds of voting, Athens finally lost out to Atlanta.

Athens' 1997 bid for the 2004 Summer Games was much more focused and detailed. In particular it addressed IOC concerns around infrastructure and air pollution in the city. The successful staging of the 1997 World Athletics Championships the month before the IOC met in Lausanne to vote on the 2004 host city also boosted Athens' chances.[2]

Buenos Aires, Capetown, Rome and Stockholm were also bidding for the Games. The voting went to five rounds, with Athens finally triumphing over Rome by a significant margin.

Greece was experiencing a massive debt-fuelled economic boom at the time, made even more frenetic by its entry into the Eurozone in 2001 after the Greek government obfuscated to conceal its parlous economic state. The Games were seen as an opportunity to regenerate the city, following the widely hailed beneficial impact of the Barcelona Games.

Ambitious plans were drawn up for a new international airport and massive investment in a new subway system, new roads, metro and tram systems, a major upgrade of all telecoms facilities and much-needed conservation work on the city's archaeological heritage. There would be no new main stadium: instead, the stadium built for the 1982 European Athletics Championships would be refurbished and upgraded.

Permanent venues for such minority sports as softball, beach volleyball and field hockey were also planned. The Greek authorities were quietly confident that the necessary funding would be available and the massive and complex construction work would be completed on time. The rest of the world was not so sure.

In the end the contractors had to import thousands of Albanian workers and institute three-shift 24-hour seven-day working. This significantly inflated costs, as did the extra security deemed essential in the aftermath of 9/11. This was estimated at 1.2bn Euros, some six times more than at the previous Summer Games in Sydney.

Nevertheless, and much to the surprise of foreign observers, everything was ready on time for the Games – just. A total of 201 countries sent 4,329 women and 6,296 men to compete in 28 different sports, and performance standards were high. One unique departure from the norm was the holding of the shot-put competition in the original stadium at Olympia, while the Panathenaic Stadium, home to the 1896 Summer Games, was the archery venue and the site of the marathon finish.

Stadium origins

Athens already possessed the 75,000-seat Spyros Louis stadium, located at the centre of a complex of sports arenas, best known by its Greek acronym AOKA, in the north-east of the city. Named after the Greek winner of the very first Olympic marathon in 1896, the stadium was designed by the German firm Weidleplan and opened in September 1982.

Precast concrete construction was used throughout. Thirty-four pairs of external precast frames, each weighing 600t, supported the stands. Only those spectators below the upper tier of seats enjoyed any protection from the elements. The most distinctive feature of the stadium were the four 62m tall inclined lighting towers, founded some distance outside the oval footprint and leaning inwards.

The stadium hosted several major athletics events and football finals before the decision was taken to upgrade it into something more iconic for the 2004 Summer Games.

Upgrade designer

Santiago Calatrava Valls (b.1951), better known simply as Santiago Calatrava, is a Spanish polymath with degrees in architecture and structural engineering. He rose to fame as one of the first designers to apply architectural aesthetics to bridge design. His first project to attract international approval was the 1987 Bac de Roda Bridge in Barcelona. Since then he has been responsible for many iconic structures, all featuring flowing, organic forms.

Upgrade design

Weather protection for the spectators was the top priority for the stadium upgrade. Calatrava's initial sketches and models featured a pair of undulating glass roofs suspended from two giant arches, requiring thousands of individual glass elements.

After several iterations the final design featured 5,000 polycarbonate glazing elements, preferred on the grounds of cost and weight. Even so, the roofs weighed in at 19,000t, and covered a total area of 25,000m².

These were supported by two "twin arches" 72m high and spanning 304m. For improved wind resistance up to 120km/hr the roof curve was steeper and higher, and the four arch bases were significantly revised. Spectator capacity was now 68,000, all seated.

The infield was upgraded with natural grass in 1.2 x 1.2m modular containers set on a concrete slab base incorporating drainage and irrigation systems. This allows the grass surface to be removed and stored outside the stadium so that different events such as concerts can be staged.

A spectacular new roof was added to the original 1982 stadium just in time for the 2004 Summer Games.
SOURCE: CREATIVE COMMONS – CREDIT SPYROSDRAKOPOULOS

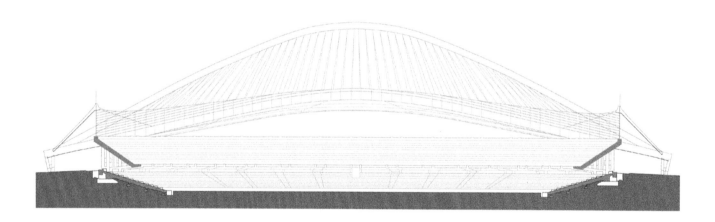

A planned 110m high Olympic Cauldron also fell foul of cost and time constraints. It would have been the tallest structure in Athens and an iconic local landmark visible from many kilometres away: what was finally built was much more modest.

One of the first key steps in the upgrade was the demolition of the four original inclined lighting towers. A series of temporary support towers was then erected on the east and west sides of the stadium to permit the assembly of the hollow steel arches.

Hydraulic rams were used to slide the completed arches 68 and 72m respectively into their final positions. It was not until the end of May 2004 that both arches were in place and ready for the start of roof installation, which was completed just in time for the Opening Ceremony.

The twin arches were assembled outside the existing stadium then slid more than 70m into their final positions.
SOURCE: CREATIVE COMMONS – CREDIT ADIRRICOR

The upgraded stadium awaits the first athletics events of the 2004 Games.
SOURCE: CREATIVE COMMONS – CREDIT MISTER NO

Legacy

Time has not been kind to most of the new Olympic venues in Greece. The continuing Greek financial crisis has meant that the necessary funds to maintain them all – estimated at £500m annually – has rarely if ever been forthcoming. Many of the venues for minority sports with virtually no local following have been left abandoned and decaying. And as the years pass it becomes increasingly obvious that the frantic rush to complete construction in time has led to shoddy workmanship and rapid deterioration.[3]

The main stadium, however, has hosted all three main local football clubs, and has held two European Cup finals. It suffers from the usual problem of a multi-sports arena: the upper seats at each end of the ground are a very long way from the football pitch. But as the average crowd for home games is around 20,000, this is rarely a problem.

Funds for maintenance are limited and inadequate. The spectacular roof leaks persistently. Nevertheless, the stadium has hosted several sell-out concerts by leading performers such as Madonna, U2 and Michael Jackson.

Athens itself did receive a much-needed upgrade of its infrastructure, albeit at grossly inflated cost. Total cost of the Athens Games is now calculated to be US$16bn,[4] making them the most expensive ever staged in terms of cost per capita.

Notes

1. Weisman, Steven R. (1990) "Atlanta selected over Athens for 1996 Olympics", *The New York Times*, 19 September.
2. Longman, Jere (1997) "Athens pins Olympic bid to world meet", *The New York Times*, 3 August.
3. Malone, Andrew (2008) "Abandoned, derelict, covered in graffiti and rubbish: what is left of Athens' £9billion Olympic 'glory'", *Daily Mail*, 18 July.
4. Goldblatt, David (2016) *The Games: A Global History of the Olympics*, New York: W.W. Norton.

CHAPTER 30

BEIJING
PEOPLE'S REPUBLIC OF CHINA, 2008

Background

Prior to the IOC meeting in Moscow in 2001 ten cities had submitted bids to host the 2008 Summer and Paralympic Games. Bangkok, Cairo, Havana, Kuala Lumpur and Seville failed to make the shortlist, leaving the field clear for Beijing, Istanbul, Osaka, Paris and Toronto.

Beijing's suitability as an Olympic venue was questioned by many: apart from the well-documented and widely condemned human rights abuses and the issue of Tibet there were serious concerns about the levels of air pollution in the city.

However, there was sympathy for Beijing following its disappointment eight years earlier, when it had led every round of voting for the 2000 Summer Games before losing to Sydney in the final round. Beijing also undertook to significantly improve air quality. In the event the city won easily in the second round of voting.[1]

Following the award, the Chinese government embarked on a massive and unprecedented capital investment programme. An estimated $40bn was spent on a vast new international airport terminal – making it the largest in the world at the time – eight new subway lines, new ring roads, a new and comprehensive sewerage system, a network of fibre optic cables and 50,000 new hotel rooms. Twelve new Olympic venues were constructed.[2]

Controversy dogged the Olympic Torch relay, at 137,000km the longest ever, passing through every continent except Antarctica and even being carried to the top of Mount Everest. On its international leg the relay encountered many anti-China, pro-Tibet protests, and the flame was even extinguished by protestors in Paris. Afterwards the IOC insisted that organizers of subsequent Summer Games confine the torch relay to their own countries.

In the end there were no boycotts and no bans. A total of 204 countries sent 4,637 women and 6,305 men to compete in 28 sports. An unprecedented number of World and Olympic records were broken. Subsequently, however, another record was set, one that cast a shadow over the legacy of the Games. In 2017, following a major investigation, 50 medal winners were stripped of their awards for doping violations.

Beijing's Opening Ceremony lasted four hours, featured more than 15,000 performers and is reputed to have cost $100m. Overall, the Games are calculated to have attracted 4.7bn television viewers, the highest on record.

In an effort to improve air quality, manufacturing sites were restricted and the number of cars on the streets reduced. Along with favourable weather conditions these measures helped to ameliorate air pollution during the Games, although it was still worse than any other Summer Games. There were no noticeable improvements in human rights, nor any change to China's position on Tibet.

Stadium origins

Beijing already had an existing National Olympic Sports Center, complete with stadium, gymnasium and aquatics centre, which was used for the 1990 Asian Games. To stage the Summer Games, however, the organizers planned on a much greater scale.

A massive relocation of residents and demolition of traditional housing extended the existing 164ha park around the National Olympic Sports Center northwards to a staggering 1,159ha, the largest Olympic Park to date. Dubbed the Olympic Green, this was intended to be an urban space for future use, forming an integral part of the city. The 680ha northernmost section of this became the Forest Park.

A 315ha central section accommodated the main Olympic venues: the main stadium, the aquatics and tennis centres, an indoor stadium and facilities for hockey, archery, fencing and shooting. The area lay on the historic north–south central axis of Beijing, north of the Forbidden City.[3]

International design competitions were held for several of the venues, particularly the main stadium and the aquatics centre. One major constraint for the designers was the city's location in one of the world's most active seismic zones, with an earthquake up to magnitude 8 on the Richter scale a serious possibility.

Design team

No fewer than 13 design proposals for the main stadium were shortlisted and submitted first to a panel of judges and then exhibited to the public. Both the professional panel and the public preferred the same entry, that proposed by a multinational team made up of China Architecture Design and Research Group (CADG), Swiss architects Herzog & de Meuron and local artist Ai Weiwei.

Project architect was Stefan Marbach of Herzog & de Meuron, while CADG was lead by Li Xinggang.

Structural design was the responsibility of UK-based Arup. Project engineers were Stephen Burrows and Tony Choi.

Design

One of the client's key requirements was a legacy use for the stadium as an all-seasons venue for field sports and athletics. Winters in Beijing can be severe, with temperatures below −20° C having been recorded. As in Montreal more than 30 years earlier (see Chapter 22), an obvious response was to opt for a retractable roof, and detail design work began on this basis.

A sliding roof design was developed, featuring two retractable sections each 70m long and spanning 75m. To minimize visual impact, the design team rejected such structural solutions as masts or arches, opting instead to wrap the roof structure closely around the elliptical central bowl.

However, to achieve the specified seismic performance it was decided to completely separate the roof structure from the central bowl. The roof structure could then be constructed as a complete entity with no movement joints, greatly simplifying the roof operating mechanisms.

Although it was soon dubbed "the Bird's Nest Stadium", the architectural design of the enveloping steel structure was inspired by such Chinese influences as the local pottery and heavily veined naturally occurring "scholars' stones". The challenge for the structural designers was to come up with a solution that was sympathetic to the architectural intent but robust enough to cope with both vertical loads and seismic events.[4]

In June 2004, after reviewing costs and progress, the client decided to drop the requirement for a retractable roof. A larger roof opening was adopted. This lightened the structure somewhat and slightly simplified the design challenges, but still required considerable creative input.

All visible structural elements, curved or straight, had to be formed from a standard 1.2m x 1.2m steel box section. Essentially, the "Bird's Nest" had three categories of components:

Swiftly dubbed "The Bird's Nest", Beijing's Olympic Stadium pushed construction technology to its limits.
SOURCE: IOC

"primary", consisting of 24 column truss structures supporting primary roof trusses spanning up to 313m, "secondary", in the outer skin of the façade only, to facilitate the cladding system, and the stairs, which were integrated into the façade. Overall height was 69.2m.

Advanced seismic analysis was performed to confirm that the steel structure could cope with earthquakes up to magnitude 8 on the Richter scale. Key components underwent prototype testing at 1:2.5 scale. In all, some 44,000t of steel was required.

The structural separation also simplified the seismic design of the central bowl. It was developed as six totally separate segments, with individual stability systems and movement joints.

Olympic stadia have rarely proved to be ideal for legacy sports, particularly soccer and American football. Spectators can be a long way from the action. One solution is a movable lower tier of seating that can be extended over the running track – a solution ultimately adopted for London 2012 (see Chapter 31).

This was not allowed for in the client's brief. An alternative solution was developed by the design team: a cantilevered middle tier of seating, with the front 15 rows of seating extending over the lower tier. Spectator experience was also improved by the choice of an elliptical planform and the reduced height of the terracing at each end. Total spectator capacity was

For maximum earthquake resistance, the stadium's concrete seating bowl is completely independent of the steel roof structure.
SOURCE: IOC

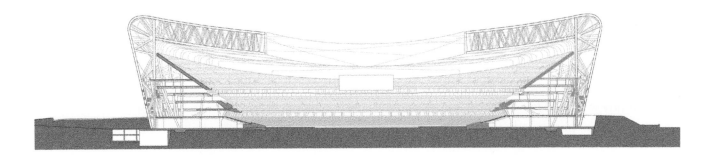

91,000, including 11,000 temporary seats. Overall the stadium measures 333m on its major axis, 298m on the minor.

Ground conditions were unchallenging. Cast in situ concrete bored piles ranging between 800mm and 1,000mm in diameter were founded in a stratum of cobbles and gravel some 36m below existing ground level. These support an in situ concrete frame made up of independent segments stabilized by column-beam action and the in situ concrete service cores.

Between the six segments are 200mm wide movement joints. Each segment measures between 120m and 150m long, with the middle and upper tiers formed from precast concrete units. The outer surface of the concrete bowl was painted red, for visual contrast, as it would be visible through the enveloping outer roof structure.

Within the steel structure are many spacious vestibules or lobbies, together with shops, restaurants and snack bars. One significant innovation was the installation of a ground source energy system beneath the stadium, capable of providing heated water in winter and cooling water in summer. Green credentials were further enhanced by a 100kW solar photovoltaic array.

At night the "Bird's Nest" reveals its intertwining structure, almost certainly the most complex of all Olympic stadia.
SOURCE: CREATIVE COMMONS – CREDIT PETER23

Legacy

In 2022 the stadium will host the Opening and Closing Ceremonies for the Winter Olympics and Winter Paralympics, making it the only stadium so far to host such ceremonies for both Summer and Winter Games.

Currently it is used for athletics, soccer matches and concerts. It continues to be a popular tourist attraction, particularly for a large number of Chinese tourists.

Notes

1. Longman, Jere (2001) "Olympics; Beijing wins bid for 2008 Olympic Games", *The New York Times*, 14 July.
2. "The cost of the Beijing Olympics", *The Guardian*. London. 28 July 2008.
3. https://www.travelchinaguide.com/attraction/beijing/olympic-park.htm
4. *Arup Journal* 1-2009.

CHAPTER 31

LONDON

UK, 2012

Background

London became the first city ever to hold three Olympic Summer Games when in 2005 it defeated bids from Moscow, New York, Madrid and Paris. In its bid, London promised to produce a Games that appealed to the youth of the world, with sport at its heart, showcasing London: a sustainable Games where permanent venues would only be built where they could be underpinned for legacy use, with a sound business plan.

It was to be the most sustainable Games ever.[1]

At the Opening Ceremony, 10,768 athletes representing 204 nations marched into the stadium. Women's boxing was included for the first time: the London Games were the first at which every sport had female competitors and to which all eligible countries sent at least one female competitor.

Origins

The big decision was to create an Olympic Park, which would house the facilities that would remain after the Games, namely the stadium, the swimming pool, the velodrome and an arena called the Copper Box (because of its cladding), which would remain as a community facility.

A challenging brownfield site was selected. This was in the Lower Lea Valley, to the east of London, surrounded by boroughs that were the poorest in the country. The area was blighted by seriously contaminated land, and occupied by a constellation of small landowners and small industries. A massive demolition, site clearance and soil cleansing programme would be needed and the River Lea would have to be cleaned up and restored.

Equally challenging was the target of more than 90 per cent by weight of all demolition material from the site to be reused or recycled.

In the interests of sustainability there would be a large number of temporary venues, mostly erected at historic locations, which would be removed or relocated after the Games. Wherever possible, existing locations, such as the O2 Arena, would also be pressed into service.

The initial brief was for an 80,000-seat stadium. It was believed there would be no need for another stadium of this capacity in London, and therefore the initial concept was to remove the upper parts of the structure after the Games. This would create a legacy stadium of reduced size, seating 25,000 spectators, to become the home for athletics in London.

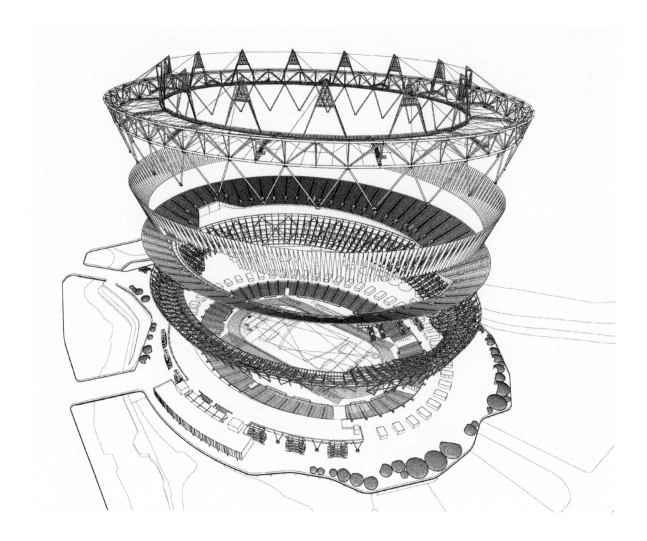

A further step taken towards a sustainable result was the Central Energy Centre, which houses the combined cooling, heat and power plant that supplies district heating to all facilities in the Olympic Park, including the stadium.

It reduces the operational carbon emissions by approximately 20 per cent, compared to a conventional approach of individual plant facilities.

A huge improvement in the transport infrastructure was essential to meet the requirements of the IOC.[2] Existing rail lines were expanded and upgraded, and a high-speed rail service introduced. Overall, the capacity of the existing network was trebled, based on the population projections for the regenerated area. There was 100 per cent spectator access by public transport, and 93 per cent of competitors could reach their events within 30 minutes.

Design team

Populous is an international architectural practice with a long track record in the design of stadia and sports facilities all over the world, including the Emirates Stadium and O2 Arena in London. It has the unique distinction of having been involved with three Olympic stadia:

Sydney 2000, London 2012 and Sochi for the 2014 Winter Games. For this project it adopted the mantra "Embrace the Temporary."

Structural designers were BuroHappold Engineering, who also designed the original Millennium Dome (the precursor of the O2 Arena) and worked on the Emirates Stadium. It too was involved in the Sochi project, and is now a major international professional services firm. It was awarded a gold medal by the Royal Academy of Engineering for the Millennium Dome project.

Design

As the detailed design process got underway, the brief changed. The stadium would not be reduced to only 25,000 seats after the Games; it would be retained as a multi-function arena with minimum modifications. It is one of the lessons that often occur, that "temporary" architecture becomes permanent.

Within the Park the stadium sits on an island bounded by the River Lea, accessible only via bridges. During the Games this had the advantage of facilitating the control of access.

Access to the stadium is via bridges across the River Lea.
SOURCE: COURTESY OF POPULOUS

The Opening Ceremony for the 2012 Summer Games gets underway.
SOURCE: COURTESY OF POPULOUS

Underlying the site are soft alluvial clays. Some 5,000 piles up to 20m deep were needed for the foundations to the base level.

The original concept was a two-tier design. A semi-sunken base tier constructed in "low-carbon"[3] concrete provided 25,000 permanent seats. Above this would be 55,000 seats in a steel and concrete upper tier, originally intended to be easily dismantled and removed after the Games.

Much of the steel used for the upper tier structure was surplus high yield steel large diameter pipes from North Sea gas pipeline projects. The seating terraces were formed of 7.5m precast concrete planks, manufactured with 30 per cent of the aggregate recycled from demolition waste concrete.

A minimalist steel structure supported the cable-stayed roof. Encircling the perimeter was a 15m high, 800m long compression truss linked by cables to an inner tension ring, the whole compared to a bicycle wheel, with the perimeter truss akin to the wheel rim, the cables acting like spokes and the tension ring mimicking the wheel spindle.[4]

Bolted connections were used, to speed its removal after the Games. The roof material was a thin membrane of phthalate-free PVC-coated polyester, which covered two-thirds of the seating.

On top of the roof structure, there was a ring of triangular lighting rigs, each 28m tall, which gave the roof an exciting and functional design feature. Computerized fluid dynamic modelling and wind tunnel testing were used to determine the format of the roof coverage, to control maximum allowable wind speeds on the track.

Changing rooms and a 60m warm-up track were located beneath the western stand. Catering facilities, retail outlets and toilets were housed in lightweight buildings on a permanent concourse surrounding the stadium. This became a lively and user-friendly space, and the buildings were designed to be capable of adaptation and change.

There were 56 entry points into the stadium from the concourse, which provided easy access.

The exterior of the stadium expressed an articulation of the external diagonal structure. This was shielded by a polyester and polyethylene wrap, acting as an outside membrane used for decoration and display.

What rose in East London was one of the lightest Olympic Stadia ever built. (It has been estimated that the steel content was 25 per cent of that in the "Bird's Nest" Stadium in Beijing, London's predecessor.)

Further justification for the lightweight design of the stadium was highlighted by the important role of embodied energy in the construction, versus operational energy, in the overall carbon footprint. This is due to the intermittent use of a stadium, in comparison with other building types.

Legacy

The design ethos of "embracing the temporary" had been adopted in the knowledge that the stadium's function was to change after the Games. It was to have an anchor tenant in the form

An external polyester and polyethylene membrane wrap around London's Olympic Stadium was used for decoration and display.
SOURCE: IOC

of a Premier League football club, and would be available for many other purposes, including the Rugby Union World Cup.

A challenging stipulation for legacy use was that the athletics running track had to be retained. To adapt the stadium for football events and the like, an automated system of retractable seating was installed, with all four sides of the lower seating bowl able to be moved out over the track. Final seating capacity is 54,000.

An extended roof was needed to cope with these new uses and cover all the seats. The fabric roof used during the Games was removed, and replaced by the largest single-span cable net roof structure in the world, 45,000m² in size, and spanning up to 84m. This required some strengthening of the original steelwork.

The new roof was designed to improve acoustics and to heighten the spectator experience, reflecting the noise from the terraces, focusing the sound and projecting it towards the central space. In order to preserve some of the Olympic Stadium's identity, the iconic triangular lighting towers that used to stand over the roof have been inverted. They now hang underneath the new larger roof.

Around the stadium the Olympic Park, now renamed the Queen Elizabeth Park, is, at 227ha, one of the largest urban parks created in Europe in the last 150 years. It was designed

After its transformation, London's Olympic Stadium played host to the 2015 Rugby World Cup.
SOURCE: CREATIVE COMMONS – CREDIT DAVID ROBERTS

to integrate habitat and landscape design into the design of the venues within it, and is acknowledged to be a major environmental achievement.

A less encouraging revelation was the announcement that, as of late 2017, 29 medal winners had been stripped of their awards for drug offences. The worst offender was Russia, with 13 medal awards rescinded.

Notes

1. https://web.archive.org/web/20091018073244/http://www.london2012.com/plans/sustainability/getting-ready/index.php
2. "Report of the IOC Evaluation Commission for the Games of the XXX Olympiad in 2012" (PDF).
3. "Low-carbon" concrete generally contains a proportion of recycled aggregates and has a significant addition of processed cementitious waste materials such as ground granulated blast-furnace slag, pulverized fuel ash or silica fume.
4. https://www.bdonline.co.uk/london-2012-olympic-stadium-by-populous/5016249.article

CHAPTER 32

RIO DE JANEIRO

BRAZIL, 2016

Background

In 2007 cities as diverse as Doha, Prague and Baku expressed an interest in staging the 2016 Summer Games, alongside Tokyo, Madrid, Chicago and Rio. Only the last four made it onto the final shortlist, with Rio eventually triumphing over Madrid in the third round of voting at the IOC meeting in Copenhagen in 2009.

At the time Brazil was still enjoying an economic boom largely unaffected by the 2008 global financial crisis, underpinned by the discovery of major offshore oil reserves. Rio felt confident enough to undertake to host not only the Summer Games but also the 2014 soccer World Cup, an ambitious commitment. There were doubts that the city would be able to deliver all the infrastructure and environmental improvements it promised, but there was a general feeling amongst IOC members that it was time for the Summer Games to be staged in South America.

This decision was made despite Rio's performance in staging the 2007 Pan American Games. These did not go well. None of the promised infrastructure and transport improvements were delivered, despite which the final cost was up to six times the original budget. Worse was to follow. Before the 2013 Confederation Cup – effectively a dress rehearsal for the World Cup the following year – there were widespread protests against the costs of the two events, protests that were met with extreme police violence.

Later there was a scare regarding the possibility of an outbreak of the mosquito-borne Zika virus amongst competitors and spectators alike. There were also complaints that the promised sewage treatment upgrade for the heavily polluted Guanabara Bay would not be ready on time for the sailing events. Overshadowing all these, however, was the economic catastrophe that engulfed Brazil following the collapse of commodity prices, particularly oil prices.

On the currency markets the Brazilian real suffered a 50 per cent devaluation. Crime and corruption flourished. By 2014 the IOC was seriously concerned at the lack of progress, declaring the situation "worse than Athens."[1] Nevertheless, and although several of the more ambitious projects were unlikely to be ready on time or were simply abandoned, there were significant improvements to the city's infrastructure.

The number of people with access to good quality public transport rose from 18 per cent in 2009, when Rio was selected as the host city, to 63 per cent in 2016. There were four new rapid transit bus lines, and a better rail service that included a new metro and tram line. The airport was improved. Large investments had also been made in upgrading cultural assets such as the buildings designed by Oscar Niemayer in Niteroi, along with tourist attractions like the Christ the Redeemer Statue in its spectacular hill-top setting, and the Sugarloaf Mountain cable car.

At the time of writing the Official Report of the Rio Summer Games had not been published. There was no final breakdown of competitor numbers by gender: currently available figures suggested that 207 countries sent more than 11,000 athletes to compete in 28 sports.[2] Around 45 per cent of the competitors were women.

For the first time ever there were ten athletes forming the Refugee Olympic Team, the IOC's reaction to the European migrant crisis. Nine "Independent Olympic Athletes" also competed: these came from Kuwait, whose Olympic Committee had been suspended by the IOC on the grounds of excessive interference by the Kuwaiti government.

Russian athletes were allowed to compete despite the World Anti-Doping Agency recommending that the country be banned following an investigation that revealed widespread state-sponsored drug abuse amongst Russian athletes.[3] In the end the IOC allowed 282 "clean" athletes to compete, while 111 were banned.[4]

There was to be no new main stadium. Rio already had two major venues, the iconic Maracanã stadium and the newer, smaller, Estádio Olímpico João Havelange. An unusual decision was taken: the Opening and Closing Ceremonies would be held in the Maracanã, but the athletics, considered by most to be the true heart of the Summer Games, would take place in the Estádio Olímpico.

However, a total of nine new and seven temporary venues were constructed. Nine of these were clustered in the Barra Olympic Park, built on the site of a former Formula 1 motor-racing circuit. Several of the temporary structures featured "nomadic architecture, being designed to be dismantled after the Games and reconfigured into pre-planned schools and aquatic centres."

Stadium origins – Maracanã

First opened in 1950 to host the 1950 soccer World Cup, the stadium became world famous when an estimated 200,000+ spectators crammed in to watch the decisive game between Brazil, the favourites, and its southern neighbour Uruguay. Brazil lost.

Over the decades the venue has seen attendances of more than 150,000 on 26 occasions, and more than 100,000 on no fewer than 284 occasions.

A design competition held in 1947 was won by a local construction company and an astonishing total of seven architects. The winning design was an oval stadium constructed of in situ reinforced concrete on a radial grid. There was a concrete roof offering limited protection to spectators. Architecturally it was seen as neither inspiring nor noteworthy.

Construction began in August 1948. Two years turned out to be not enough time to complete the stadium before the 1950 World Cup; nevertheless the authorities allowed the competition to go ahead even with scaffolding inside the ground. It was another 15 years before the stadium was finally complete.

In 1992, following the collapse of part of an upper stand which left three spectators dead, the decision was taken to convert the stadium to an all-seated venue. This significantly reduced spectator capacity; after further renovations in 2005–2006 this totalled 87,000.

Although widely and popularly known as the Maracanã, the stadium's official name since 1966 is Estádio Jornalista Mário Rodrigues Filho, named after a prominent advocate of the stadium who died in that year.

Olympic upgrade – Maracanã

In 2010 a significant renovation project was launched to bring the stadium up to the required standards for the 2014 World Cup and the 2016 Summer Games. As the stadium had been classified as a national landmark in 1998, there was a need for sympathetic reconstruction.

Design team

Designer for the roof was Schlaich Bergermann Partner of Germany. The Brazilian architects for the seating bowl and the remainder of the stadium were Fernandes Arquitetos Associados.

Design

To respect the original design and protect the cultural heritage it was decided to maintain the façade while upgrading the interior and adding a much more extensive yet unobtrusive roof. This constrained the architects to work within a confined structural envelope, but left the view from outside almost unchanged.

Originally the concrete seating bowl had two tiers of spectator accommodation. This was to be changed to a single tier layout by demolishing the lower tier and extending the upper tier down to arena level. The lower section was formed of precast concrete sections supported by structural steelwork.

An internationally used tread depth was adopted for more comfortable seating. Final seating capacity was 78,838, 95 per cent of which was protected by a new roof that replaced the original, much smaller roof.

An elegant lightweight canopy that appears to float above the seating bowl, the new roof is supported off original concrete roof columns. Conceptually it resembles the London 2012 roof (see Chapter 31). The steel supporting structure is based on the principle of a horizontal spoked wheel with three central tension rings and a perimeter compression truss. Roof material is a tensioned polytetrafluoroethylene (PTFE) coated fibreglass membrane. Four large video screens are suspended from the roof.

Stadium origins – Estádio Olímpico

A dual stadium approach was first adopted for the 2007 Pan American Games. The Maracanã would host the Opening and Closing Ceremonies, together with some soccer matches.

Rio's iconic Maracanã stadium was upgraded with a London 2012-style roof and used for the Opening and Closing Ceremonies.
SOURCE: CREATIVE COMMONS: CREDIT DANIEL BASIL

However, a brand new stadium would be built for athletics and soccer.

This was to be named after former Federation Internationale de Football Association (FIFA) chairman and IOC member João Havelange – who died at the age of 100 during the 2016 Summer Games.

Design team

Local conglomerate Odebrecht S.A headed up a consortium of contractors that won the design and build contract for the stadium. Lead architect was Carlos Porto, heading up an in-house team of engineers and architects

Design

Reinforced concrete was the structural material of choice for the main stadium bowl. Four massive white tubular steel arch trusses and cables support the metal roof and define the architectural form. Like the earlier Maracanã, the architecture attracted few plaudits.

Construction began in 2003, and dragged on until just before the opening of the Pan American Games. Later it was estimated to have run 533 per cent over budget.[5] Spectator capacity was 46,931.

Athletics events were held in the Estádio Olímpico.
SOURCE: CREATIVE COMMONS: CREDIT GABRIEL HEUSI

After the Pan American Games the stadium was rented by local soccer club Botafogo. Eight days after the deal was signed a 15m long by 6m high section of concrete wall collapsed, fortunately without injuring anyone. In 2013 fears over the structural safety of the roof in high winds led to a two-year closure for strengthening work.[6]

For the 2016 Summer Games temporary seating brought spectator capacity up to 60,000. An unusual and somewhat controversial blue running track was installed.

In 2017 the stadium was officially renamed as Estádio Olímpico Nilton Santos, in honour of a Brazilian football player regarded as one of the greatest defenders of all time.

Legacy

It has become increasingly clear that the preparations for both the 2014 World Cup and the 2016 Summer Games were mired in corruption at all levels. National and local politicians and officials were named, shamed and, in many cases, indicted. The political fallout from the ongoing corruption investigations is still hard to predict at the time of writing.

As a result, many of the Games venues seem to have languished in an administrative limbo. The Maracanã in particular attracted critical media attention: the playing surface became rutted and withered, seats were ripped out, equipment stolen. No one would accept responsibility for

Rio's Estádio Olímpico featured an unusual blue running track.

its management and maintenance until April 2017, when the multinational media conglomerate Lagardère took over.

Local soccer club Botafogo continues to play at the Estádio Olímpico, which has undergone further upgrades since the Summer Games. It also hosts major concerts – although not as many as the Maracanã.

At the time of writing it was difficult to evaluate the real legacy of the Rio Summer Games. There were some obvious infrastructure and environmental improvements, but it was still too early to put these into context.

Notes

1. "Rio's Olympic preparations 'worst' ever, says IOC's Coate", Reuters, 29 April 2014.
2. https://web.archive.org/web/20160821031645/https://www.rio2016.com/en/athletes
3. Ruiz, Rebecca (2016) "Russia may face Olympics ban as doping scheme is confirmed", *The New York Times*,18 July.
4. "IOC confirm 278 Russian athletes are eligible to compete at Rio", *Daily Mail,* 9 July 2016.
5. Bandeira, Luiza (2007) "Clube Botafogo administrará estádio olímpico do Engenhão", *Agência Brasil* (in Portuguese), 3 August.
6. https://www.insidethegames.biz/articles/1014572/rio-2016-athletics-venue-to-be-closed-until-year-before-games-while-urgent-repairs-carried-out

CHAPTER 33

TOKYO
JAPAN, 2020
(DUE TO TAKE PLACE 24 JULY TO 9 AUGUST)

Background

In 1964 Tokyo became the first Asian city to host the Summer Games. When the Summer Games return to the city in 2020 it will also become the first Asian city to host the Summer Games twice.

Tokyo was up against Istanbul and Madrid when the IOC met in Buenos Aires in 2013, eventually beating Istanbul by 60 votes to 36 in the final round following Madrid's elimination. The city's bid had been strengthened by its experience in unsuccessfully bidding for the 2016 Summer Games, where it trailed behind Rio de Janeiro and Madrid in the first two rounds of voting before being eliminated.

After the 2013 vote, IOC President Jacques Rogge commented: "We are confident that our Japanese friends will be able to provide excellent Games." The accolade "safe pair of hands" was used several times in the bidding process. "Innovation, using Japan's renowned creativity and technology to benefit sport & the Games" were also words used.[1]

Japan's economic strength, mature democracy and political stability offered strong foundations for the Games. Initially there were concerns over contaminated water: there was an earthquake and tsunami in 2011, and there had been a disaster at the Fukushima Daiichi Nuclear Power Plant. There had been subsequent claims about levels of radiation. Assurances were given to the IOC that these would cause no problems.

Japanese athletes first competed in the 1912 Games in Stockholm. In 1930 Baron Pierre de Coubertin wrote: "Last year in Geneva, one of the Japanese delegates at the League of Nations said to me 'it is impossible to imagine to what extent the revival of the Olympic games has transformed my country.'"[2]

For the Japanese capital the 1964 Summer Games was a landmark event: it was instrumental in the city's economic development and social reconstruction. The international architectural impact was enormous (see Chapter 19). Tokyo 2020 promises to be equally seminal.

A highly compact design concept puts the sport and spirit of the Games in the heart of the city. About 85 per cent of the Olympic venues should be within an 8km radius of the athlete's village.

Stadium origins.

There had been talk of upgrading or reconstructing the 1964 Olympic Stadium, but it was decided that demolition was the better course, with a new stadium to be built on the same site. Demolition of the old stadium was completed in October 2015.

This was in fact the second major stadium to occupy this site, centrally located in the Tokyo Meiji Park. In 1924 the 35,000-spectator capacity Meiji Jingu Gaien Stadium opened and was due to be used for the cancelled 1940 Summer Games. It was later upgraded to 65,000 seats. This was demolished in 1957 to be replaced by a new National Stadium with a capacity of 58,000, mostly seated. For the 1964 Games this was upgraded and extended, with a final capacity of 71,328.

Design team

An international design competition was held in 2012 with the jury chaired by the eminent Japanese architect, Tadao Ando. The winning design was by the Iraqi-born British architect Zaha Hadid. Zaha Hadid had an imposing reputation: she had won the Pritzker Prize, the Stirling Prize twice and was awarded the Royal Gold Medal by the Royal Institute of British Architects, and had designed the Aquatics Building for the London 2012 Olympics.

Her futuristic design for the stadium, with a sweeping curved structure, proved to be highly controversial. It was claimed that there was public concern over the costs of the design. There had also been concern expressed by prominent Japanese architects, including Fumihiko Maki and Toyo Ito, over the suitability of the design for the site, and for Japan. As a result, the Japanese Prime Minister Shinzo Abe scrapped this design in July 2015. The plans for a retractable roof over the stadium were also abandoned.

Tragically, Zaha Hadid died in April 2016, at the age of 65.

In December 2015 the Japan Sports Council announced that the Taisei Corporation, Azusa Sekkei Co. Ltd and Kengo Kuma and Associates Joint Venture had been selected to design and construct the new Olympic stadium, with a new reduced cost. Inevitably, there was now a shorter programme.

Kengo Kuma is a renowned Japanese architect. In 1997 he won the Architectural Institute of Japan Award. His stated goal is to recover the tradition of Japanese buildings, and to reintegrate these traditions for the 21st century.

Design

Kuma's design is for an oval structure, with a huge oculus over the central track and field arena, featuring a wood-and-steel composite roof.

Standing up to 50m high or less, the stadium has the central arena set below ground level. Circulation and concourse areas around the stadium will feature plants and trees to respect and echo the greenery of the surrounding park. Its seating capacity will initially be 60,000 in athletics mode with the option to upgrade this to 80,000 seats in the future.

Computer renderings showing conceptual images of the new stadium at completion. These may be subject to change.
The greenery is a projection of approximately 10 years after completion.

SOURCE: COPYRIGHT DESIGN WORKS AND CONSTRUCTION WORKS OF TAISEI CORPORATION, AZUSA SEKKEI CO. LTD AND KENGO KUMA AND ASSOCIATES JV / COURTESY OF JSC

The new stadium under construction in November 2018.
SOURCE: CREATIVE COMMONS

For the Paralympics some of these seats will be replaced with 747 disabled/wheelchair accessible seating. These figures are correct as of October 2018, but have yet to be finalized.

In search of construction efficiency, precast and semi-assembled components feature in the construction, which began in December 2016 and was set to be completed in November 2019. This will not be in time to host the 2019 Rugby World Cup, which will be held at another venue. During the 2020 Games the stadium will host the Opening and Closing Ceremonies, the track and field competitions, and the football.

Legacy
An integral part of the Tokyo legacy plan is the formation of the Olympic Legacy Commission. A positive physical legacy as a result of the Games is expected.

Notes
1. *Japan Times*, 8 September 2013.
2. "Vision, Legacy and Communication", Tokyo 2020.

CHAPTER 34

THE OLYMPIC STADIUM OF THE FUTURE

Legacy and location

A primary consideration for the Olympic Stadium must be the legacy. What happens to the facility when the events are over must be a priority.

The stadium should be a part of regeneration for the city. London 2012 is a key example of this, bringing new life to a neglected part of the city. The Barcelona Olympics of 1992 catalysed a massive transformation in that city's decaying urban environment.

A number of factors have to be taken into account. Because of its importance, the design can achieve "iconic" status. It is desirable that it will have a high standard of architecture. This was an ambition of Baron de Coubertin, the founder of the modern Olympic movement. He wanted to bring "culture" into its architecture (see Chapter 3). However, the stadium must not be a "white elephant". Its usage must be realistic for the long term, a benefit to the city and its community.

Stadia of the future should reflect the heritage of the Olympic movement, and the city location. However, there is a deeper change that may be of importance. In the Middle Ages and later, before the advent of the modern stadium, the centre of the city was the place where events were held, usually in a central square. Temporary seating would be erected. This tradition continues in cities like Sienna, Florence and Pamplona. Now the trend is for the stadium to return to the heart of the city.

Stadia are buildings that have a significant impact on cities and on people's lives. They are places where memories are formed that are outside the everyday experience. Weaving a massive venue like an Olympic stadium into the fabric of a city is exciting, as is constantly negotiating with the identify of a location.

The trend is towards mixed-use developments, integrating peripheral land use. Connectivity to the urban fabric should not be an option, it should be a necessity.

Olympic Agenda 20/20

All new proposals for Olympic stadia will need to take account of the recommendations of the Olympic Agenda 20/20. The following are some of the main points.

- There will be a special focus on legacy and sustainability. A sustainability strategy should be developed, to encompass social, economic and environmental spheres. This will include operations.

- Sport for people with different abilities is important.
- The use of temporary and demountable facilities should be examined. In the case of stadia (as with Sydney 2000) the capacity can be increased for the event by the use of temporary seating, which can be removed if there is no legacy justification.
- The size of the Olympics should not be increased. Flexibility to allow management to use the stadium in different ways will be important.
- The blending of sport and culture should be strengthened. (This has relevance to the architecture of the stadium.)

Inclusivity

All future Olympic stadia should be designed on the basis of inclusivity. This means catering for all aspects of society, and excluding none.

> *It is time to challenge the polarised separation of "able-bodied" and "disabled" and realise that we just need to design for people. Good design is inclusive – it makes places that everyone can use.*
>
> (Jane Duncan, RIBA President 2015–2017)

Design can remove the frustrations experienced by many people with disabilities. It is in the hands of the designer to eliminate the barriers that create undue effort and separation. Everyone should be able to use buildings equally and independently. This is the vision of inclusivity.

There are two main features:

- Catering for the spectator. This will mean freedom of movement and access, toilet and refreshment provision and some choice in places for viewing the arena, with good sightlines.
- Catering for the sports people and the performers in the arts and entertainment who will appear in the arena, including their coaches and support staff. The requirements of the equipment of sports people have been growing in size and technology.

The problems of adapting and modernizing existing facilities have to be faced and dealt with. This is often not easy to achieve. An example not related to stadia might be worth quoting. The ruins of the ancient city of Pompeii obviously presented a barrier to inclusivity, particularly for those in wheelchairs. The pavement areas have been provided with a level surface due to an enlightened policy of providing access to all.

Increasing spectator demands

The demands of the public who attend stadia events are continually increasing. More comfortable seats are expected, along with wider access.

The following statistic based on a recent survey (Oracle Hospitality: The Fan Experience) is interesting. It is not relevant so much to Olympic events, but to other sporting occasions. Of

Vision of the future from National Geographic Magazine

SOURCE: COURTESY OF NATIONAL GEOGRAPHIC MAGAZINE;

those who attend, 51 per cent are there to support an individual team or club, but 49 per cent indicated that attending sporting events is a fun social activity with friends or family.

This means that the stadium will need to cater for supporting events and facilities in the circulation arenas.

The phrase "watching but not watching" has been used to explain this phenomenon of many spectators being there for social reasons.

Sustainability

Designing for sustainability is vital, given the concerns about the effects of global warming, the shortage of renewable resources and the future of our planet. Care over the environmental effects of what we do is essential. Stadia can be leaders in this because of their high profile. The environment is the third platform of the Olympic movement, following sport and culture.

Sustainable development can simultaneously further economic, social and environmental goals, giving importance to all.

The stadium is the largest and most visible venue facility of the Olympic games. It provides

an opportunity to illustrate the importance of sustainable and environmentally conscious planning, design and construction.

Few stadia are in full-time use, thus their energy needs for heating, lighting, cooling and essential equipment are relatively low. Their embodied energy, that was expended on the extraction, processing and transportation of the materials used for construction, is therefore relatively high. Careful selection of materials, the use of significant proportions of recycled materials, sourcing materials as nearby as possible and the minimization of road transport will all help to reduce embedded energy and associated carbon dioxide emissions.

Consideration of embodied energy means that every facility should be used as much as possible. The design must incorporate the need for flexibility to allow for multi-purpose use.

The embracing of temporary facilities should be considered. This can refer to increases in spectator capacity and facilities, which are not needed for legacy use, and can be removed after the event (the stadium built for the Sydney 2000 Olympics is an example).

The use of natural ventilation and maximizing natural daylight will reduce the amount of energy needed for air-conditioning, artificial ventilation and artificial lighting.

Renewable energy is now a mainstream way of generating part or all of the energy needs of stadia. Renewable energy sources are those that are naturally replenished. The four main types

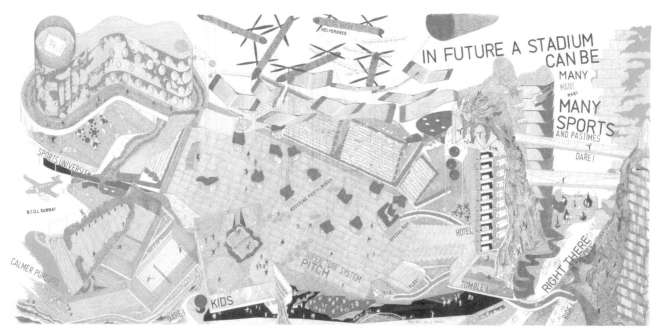

Vision of the future from Architect Sir Peter Cook

SOURCE: COURTESY OF POPULOUS;

are: solar, wind, ground, water and air source energy, and biomass.

Photovoltaic cell (PV) panels for electrical energy production and solar hot water panels should be integrated into the roof design where possible. Achieving the optimum orientation for such panels can have a major impact on the roof design.

Water is by far the greatest mass of potential recycled content. Depending on the climate of the facility, this can be used for watering the field of play, and for cleaning and flushing toilets. Waste water from showers and washbasins can be reused for the flushing of WCs.

A well-insulated basic fabric is essential. This helps to minimize heat loss in the winter and overheating in the summer.

Good air tightness with a heat exchange system will allow a proper intake of fresh air, without losing control of the temperature.

Solar protection: shading for large areas of glazing that face the sun will avoid overheating and subsequent demands on any cooling systems.

The location of trees of deciduous species can also shade the facility from sunlight during summer months and then allow sunlight to penetrate the facility in the winter months. Geographic location will be a major determinant of this.

The use of efficient service systems will reduce the operational energy used during the lifetime of the stadium.

As well as being an efficient building in its own right, the stadium can form part of a wider network of urban systems. It can benefit from the environment within which it is located.

Forming part of an eco-system can play on the unique aspects of a stadium, e.g. there is the potential for harvesting and creating energy. This will help balance the need to use large amounts of energy on event days.

Achieving sustainability in stadia in very hot climates is not easy because of the need for cooling and the energy involved, but there is research well advanced into how this could be achieved.

Multi-purpose use

It is vital for the stadium to support a full life for the community. It must not be a "white elephant" used only for occasional events. This makes sense economically, for sustainability and from the point of view of benefitting the community it serves.

Multi-functionality and transformable spaces are a necessity. So the provision and design of supporting accommodation must provide the means for full use and a variety of possibilities for the management/operators.

There is an endless list of ways to achieve this, dependent on the location and local and national circumstances. Examples include:

- Refreshment areas and restaurants with the capability of use every day of the week.
- Facilities for conferences, exhibitions and social events.
- Visitor facilities to allow for tours and a possible museum. It is possible for the stadium to become a major object of tourism. (It is not an Olympic stadium, but the Camp Nou Stadium in Barcelona attracts nearly two million tourist visitors per annum. The Olympic stadium in Beijing is also a major tourist destination.)
- Links to an adjacent arena.

Only one Olympic stadium actually contained a hotel, Moscow's Luzhniki Stadium, home of the 1980 Summer Games (see Chapter 23). However, there are examples in other major stadia. Where such a facility is impractical, the advantages of siting the stadium to link with adjacent hotel and commercial facilities are obvious.

Embracing technology

The stadium will need to incorporate and take advantage of the ever-increasing capabilities of modern technology. At each Summer Games the giant screens inside the stadium bowl are larger and of higher definition than four years earlier. They should be easily viewed by spectators from all parts of the stadium.

For big events where the demand for tickets is vastly larger than the capacity, there can be screens outside the stadium to enable the event to be watched by a larger audience.

- There can now be individual screens at the stadium seats.
- There can be controls for ordering refreshments from the seat.
- Wi-Fi and other communication opportunities have arrived in stadia.

Safety and security

Disasters at sports venues are sadly not unknown. Thankfully they are rare events: nevertheless, the need for safety and security must be taken into account in their design, construction and operations.

Design is important to create a safe legacy for the city: not a fortress, but a safe place of delight. Crowd behaviour is one of the main risk factors, along with crime, terrorism and natural events such as earthquakes, floods and extreme weather.

There are several measures that can reduce the risk of terrorist attacks. These include:

- Controlled access through the perimeter of the stadium.
- Vehicle parking and access to be kept to the perimeter of the stadium, to prevent vehicles filled with explosives being parked close to the structure. This can be achieved by clever landscaping of the facility.
- However, key vehicles such as ambulances, fire engines and police cars must have unhindered access.
- Places of concealment for bombs and chemical attack materials must be designed out. Examinations of the structure during the construction phase must take place.
- Air intakes to the ventilation system of the stadium must be designed for locations out of reach of those who might wish to use them for a chemical attack.
  More general precautionary measures include:
- The structure must be able to stay standing even if part of it has been demolished by explosion, natural disaster – or even impact from a heavy goods vehicle, accidental or otherwise. The possibility of progressive or disproportionate collapse must be eliminated at the design stage.
- Control rooms and command facilities must be able to function throughout an emergency, with independent power supplies and broadcast ability. A separate back-up location should be considered, to cope with the situation where the primary control and command facility might have been destroyed or damaged.

Clever design can ensure that none of these safety features affect the creation of a safe stadium that is a delight for the user.

Overlay

Overlay is the temporary elements that need to be added to more permanent buildings to create spaces that are needed for particular events or festivals.

A design-inclusive approach is desirable: by considering the cultural and design aspects alongside the commercial, function and safety requirements, the user satisfaction and experience can be maximized. Such an approach can create memorable places and spaces, often out of simple, tried and trusted commodities.

This is an enormously important aspect for big events like the Olympic Games, and it should not be underestimated. Included are the large spaces needed for TV and the media, the

Vision of the future from Populous

SOURCE: COURTESY OF POPULOUS

warm-up facilities for each sports space, the security facilities, and the feeding of athletes and officials. There might also be what are called "Fan-Zones", created for the event.

In terms of the stadium, this results in large temporary facilities mostly outside and around the structure.

The aim should be to treat temporary places with the same attention and standards that would be applied to permanent equivalents. Overlay should be an integral part of the planning for the event from the start, embracing the temporary, and making it an integral part of the masterplan. Every venue overlay can offer an enhanced experience to every user, and leave a lasting memory.

Opening and Closing Ceremonies

These spectacular events have become increasingly complex and expensive, as each successive Games strives to surpass its predecessors. It is these events that attract the largest global TV audiences, and it is the demands on the main stadium design that result from the creative inspirations of their planners that can be the biggest challenge for the stadium design team.

At London 2012 the roof had to be designed to carry an extra 1,000t of loading from access platforms, extra lighting and the like, while also providing accommodation within the roof space for acrobats and performers. Such demands are unlikely ever to be repeated: the challenge is to avoid creating permanent infrastructure that will be surplus to any foreseeable

How Opening Ceremonies have changed – contrast (top) Antwerp 1920 with (bottom) London 2012

legacy use. Temporary provisions are to be preferred, but this will need proper planning and design. Early discussions with the ceremony teams are essential.

Key factors include:

- Consideration of the overlay. There needs to be space for 15,000 or more athletes and officials to muster before the ceremony, plus a separate muster space and access for the usual large number of performers.
- Service roads and stadium access points will have to be large enough to cater for all ceremony equipment and structures. Again, these are more than likely to become larger and heavier.
- Stadium energy demands will peak during these ceremonies and are unlikely to reach such levels ever again. Nevertheless, adequate power cabling must be installed.
- Finally, provisions must be made for the lighting of the Olympic Flame, including access.

Stadia of the future are likely to be leaner and greener, stripped down to their essential features, cheaper to build and operate, easier to convert into a realistic and economic legacy use. Structural timber is likely to be a more popular option. Perhaps there will finally be an acceptance of the basic mismatch between an essentially oval athletics stadium and a rectangular pitch-based popular sport such as soccer or American football.

It is even possible that the normal and natural impulse of Summer Games organizers to create an iconic landmark main stadium that will attract global admiration will be modulated by more sober voices. Whatever the stadium might be – a glittering showcase for the best in modern architecture and structural engineering or a tired old workhorse given a lick of paint and a few spangles – one thing is certain: for those who compete in it and those in the audience, the Olympic Stadium will always be the "Theatre of Dreams."

SOCHI WINTER GAMES
RUSSIAN FEDERATION, 2014

Background

Until the Sochi Games there had been only one stadium that in any way resembled the landmark iconic main stadia that are usually the focus of any Summer Games. This was the Stadium Olympico in Turin, Italy, which hosted the Opening and Closing Ceremonies of the 2006 Winter Games.

However, the origins of the Turin stadium dated back to 1933. Upgrading it to modern standards cost 30m Euros, and after the Games it became the home ground for both local football clubs and a successful concert venue. What was to rise on the subtropical shores of the Black Sea was something quite different.

Until 2007 Sochi was virtually unknown to the outside world. It had been a favourite summer holiday resort for the Soviet elite, and continued to flourish modestly after the collapse of Communism. There was a small and even less well-known ski resort in the mountains to the east, with little in the way of modern infrastructure.

For political reasons both internal and external President Putin lobbied hard to win the right to stage the Winter Games in Sochi and committed massive funding to transform much of the region. There were new road and rail links, extended airport terminals and runways, a terminal for cruise ships, vastly increased power generation and telecommunications capacities, a new sewerage system and many hotels.

An Olympic Park was built on the coast at Sochi and the existing ski resort in the mountains was transformed. It is estimated that the total cost of staging the Winter Games topped $50bn, making them the most expensive Games, Summer or Winter, ever staged. The main stadium, officially known as the Fisht Olympic Stadium, seated 40,000 for the Opening and Closing Ceremonies. It cost $779m.

Main stadium design team

MosProek4 headed up the architectural team from London-based Populous.

Structural design was the responsibility of BuroHappold Engineering.

The closed roof installed for the 2014 Winter Olympics was later removed.
SOURCE: CREATIVE COMMONS – CREDIT SKAS

Design

To the south of the stadium the shores of the Black Sea are less than 200m away, whilst to the north directly on the stadium central axis is the Olympic Cauldron and Medals Plaza. This beautiful site, full of the grandeur of nature and the exciting potential of capturing forever the essence of the Olympic spirit, fundamentally inspired the stadium design.

An elevated main concourse was created, reached by monumental sweeping staircases and ramps that enabled a fully elevated circumnavigation of the stadium. The stadium user thereby enjoys constantly changing views alternatively of the snow-capped mountains and of the Black Sea.

The concourse creates a podium, like the tops of sand dunes nearby, on which rest the sea-inspired asymmetrical shells of the stadium roof, covering the equally asymmetrical/tilting structural frame like translucent crustaceans' shells.

To the north and to the south the ends of the stadium are designed to be lower than the sides, opening up views out to the sea and mountains. Also, the upper concourses are not enclosed except by the cover of the roof shell.

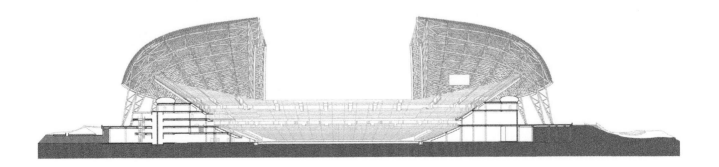

In colours ranging from cool greens at the bottom to dark blues at the top the stadium seats reflect the adjacent layering of the sea, and thus ensure that the design response to the site continues right into the details.

Sochi lies on the north-eastern shores of the Black Sea, an area characterized by a high seismicity. An assessment of the potential for liquefaction of soils during a major earthquake was assessed as possible to depths of 6m below the site.

In response, the team developed a stadium design made up of two distinct elements: the 36,500m² ethylene tetrafluoroethylene (ETFE) roof and its supporting steel structure and the in situ concrete seating bowl. The roof structure is made up of two sections covering the east and west stands, each supported along its front edge by a 270m long, 65m high primary arch running north/south.

Each arch is trapezoidal in cross-section, asymmetric on elevation and inclined to the vertical. Lateral stability is provided by a series of secondary trussed frames running transverse to the roof. Primary arch restraint is provided by the ground via piled abutments.

The Fisht Stadium separate steel and concrete elements are clearly revealed in this construction image.
SOURCE: COURTESY OF BUROHAPPOLD

Secondary frames sit directly on the reinforced concrete podium structure and also provide vertical support to the tertiary trusses that support the cladding system. To ensure that the roof is independent of the bowl under seismic loading this portion of the podium structure is fully structurally independent of the bowl structure under horizontal load.

In situ reinforced concrete makes up most of the bowl structure, along with precast concrete terrace units and steel frames to the demountable seating zones. Movement joints divide the bowl structure into eight structurally independent zones, each stabilized by a series of in situ concrete shear walls.

Floors are typically of in situ concrete beam and slab construction and act as rigid diaphragms. The lower bowl in each of the eight zones is also structurally independent under lateral loads, as it will have a significantly different response to a seismic event than the rest of the bowl.

Legacy

To comply with FIFA's regulations, $46m was spent on renovating the stadium in preparation for hosting some of the matches in the 2017 FIFA Confederations Cup and 2018 World Cup. This involved the removal of the original closed roof.

Acknowledgements

With thanks to Damon Lavelle of Populous and BuroHappold Engineering's Matthew Birchall.

After the roof was removed, and with no athletics track, the stadium was easily converted to an ideal football venue.

APPENDIX B

INTERCALATED GAMES

WHEN THE IOC met in Paris in 1901 the debacle of the previous year's Summer Games was still fresh in their minds (see Chapter 6). The contrast with five years earlier, when the Greek authorities had organized an Olympics that was much closer to the original Olympian vision of de Coubertin and his allies, must have been galling.

Athen's gleaming marble stadium packed with enthusiastic spectators, the smooth ·organization of the Games and the widespread popular support in Greece vastly strengthened Athen's bid to host the Summer Games in perpetuity.

Despite the pressure, de Coubertin was determined not to abandon his original vision of an internationalist Olympics that would ultimately involve most of the world's advanced nations. A compromise was reached. Every four years a Games would be held in a different country. In between would be a Games held always in Athens – these would be known as Olympic Games in Athens.

Following on from the low key and only marginally international 1904 Games in St Louis (see Chapter 7), what are now known as the 1906 Intercalated Games were hailed as another Greek triumph. Twenty nations sent more than 800 athletes competed in 13 sports. For the first time in Olympic history the Opening Ceremony saw individual teams march into the Panathenaic Stadium behind their national flags. National flags of the victors were raised for the first time, and the first formal Closing Ceremony was held in front of a large crowd.

However, plans for subsequent Games fell foul of geopolitics and the dawning realization that two years between Games was not enough. After the First World War the concept was quietly abandoned. The IOC later rescinded the 1906 Game's original classification as an Olympic event, and relabelled it as an Intercalated Games. Medals awarded in 1906 are no longer accepted as genuine Olympic medals.

Nevertheless, the influence of the 1906 Games on Olympic history is profound. Apart from establishing some high-profile Olympic traditions, its tight timescale and compact venues were a blueprint for all future Summer Games.

INDEX